Mathematics for Industry in Denmark: The ESGI47 and Mathematics for Industry Workshop

Journal of Physics: Conference Series Volume 52

Sonderborg, Denmark
24 – 29 August 2003

ISBN: 978-1-61738-262-8

Printed from e-media with permission by:

Curran Associates, Inc.
57 Morehouse Lane
Red Hook, NY 12571
www.proceedings.com

Some format issues inherent in the e-media version may also appear in this print version.

Printed by Curran Associates, Inc. (2010)

Institute of Physics
Dirac House, Temple Back
Bristol BS1 6BE UK

Phone: 44 1 17 929 7481
Fax: 44 1 17 920 0979

techtracking@iop.org

TABLE OF CONTENTS

Methods of Mathematical and Computational Physics for Industry, Science, and Technology

Roderick V N Melnik *et al* 2006 *J. Phys.: Conf. Ser.* **52** doi: 10.1088/1742-6596/52/1/E01 (Help)

Roderick V N Melnik and Frands Voss

Many industrial problems provide scientists with important and challenging problems that need to be solved today rather than tomorrow. The key role of mathematical physics, modelling, and computational methodologies in addressing such problems continues to increase. Science has never been exogenous to applied research. Gigantic ships and steam engines, repeating catapult of Dionysius and the Antikythera `computer' invented around 80BC are just a few examples demonstrating a profound link between theoretical and applied science in the ancient world. Nowadays, many industrial problems are typically approached by groups of researchers who are working as a team bringing their expertise to the success of the entire enterprise. Since the late 1960s several groups of European mathematicians and scientists have started organizing regular meetings, seeking new challenges from industry and contributing to the solution of important industrial problems. In particular, this often took the format of week-long workshops originally initiated by the Oxford Study Groups with Industry in 1968. Such workshops are now held in many European countries (typically under the auspices of the European Study Groups with Industry—ESGI), as well as in Australia, Canada, the United States, and other countries around the world. Problems given by industrial partners are sometimes very difficult to complete within a week. However, during a week of brainstorming activities these problems inevitably stimulate developing fruitful new ideas, new approaches, and new collaborations. At the same time, there are cases where as soon as the problem is formulated mathematically, it is relatively easy to solve. Hence, putting the industrial problem into a mathematical framework, based on physical laws, often provides a key element to the success. In addition to this important first step, the value in such cases is the real, practical applicability of the results obtained for an industrial partner who presents the problem. Under both outlined scenarios, scientists and mathematicians are provided with an opportunity to challenge themselves with real-world problems and to work together in a team on important industrial issues.

This issue is a result of selected contributions by participants of the meeting that took place in the Sønderborg area of Denmark, one of the most important centers for information technology, telecommunication and electronics in the country. The meeting was hosted by the University of Southern Denmark in a picturesque area of Southern Jutland. It brought together about 65 participants, among whom were professional mathematicians, engineers, physicists, and industrial participants. The meeting was a truly international one, with delegates from four major Danish Universities, the UK, Norway, Italy, Czech Republic, Turkey, China, Germany, Latvia, Canada, the United States, and Finland.

Five challenging projects were presented by leading industrial companies, including Grundfos, Danfoss Industrial Control, Unisensor, and Danfoss Flow Division (now Siemens). The meeting featured also the Mathematics for Industry Workshop with several distinguished international speakers. This volume of *Journal of Physics: Conference Series* on `Methods of Mathematical and Computational Physics for Industry, Science, and Technology' contains contributions from some of the participants of the workshop as well as the papers produced as a result of collaborative efforts with the above mentioned industrial companies. We would like to thank all authors and participants for their contributions and for bearing with us during the review process and preparation of this issue. We thank also all our referees for their timely and detailed reports. The publication of the proceedings of this meeting in Denmark was delayed due to problems with a previous publisher. We are very grateful that *Journal of Physics: Conference Series* kindly agreed to publish the proceedings rapidly at this late stage.

As industrial problems become increasingly multidisciplinary, their successful solutions are often contingent on effective collaborative efforts between scientists, mathematicians, industrialists, and engineers. This volume has provided several examples of such collaborative efforts in the context of real-world industrial problems along with the analysis of important physics-based mathematical models applicable in a range of industrial contexts.

Roderick V N Melnik, Professor of Mathematical Modelling, Syddansk Universitet (Denmark) and Professor and Canada Research Chair, Wilfrid Laurier University, Waterloo, Canada E-mail: rmelnik@wlu.ca

Frands Voss, Director of the Mads Clausen Institute, Syddansk Universitet (Denmark)

Institute of Physics Publishing
doi:10.1088/1742-6596/52/1/001

Journal of Physics: Conference Series **52** (2006) 1–13
Mathematics for Industry in Denmark

Flow-Acoustic Properties of Elliptic-Cylinder Waveguides and Enclosures

Morten Willatzen[1], **Lok C. Lew Yan Voon**[2]

[1]Mads Clausen Institute for Product Innovation, University of Southern Denmark, Grundtvigs Alle 150, DK-6400 Sønderborg, Denmark

[2]Department of Physics, Worcester Polytechnic Institute, 100 Institute Road, Worcester, Massachusetts 01609, USA

E-mail: `willatzen@mci.sdu.dk`

Abstract. The acoustic pressure in elliptic-cylinder waveguides and enclosures formed by a finite-length elliptic cylinder and two plane surfaces ($z = z_1$ and $z = z_2$) is discussed. The influence of a medium mean flow v_z along the axial direction on the acoustic pressure in elliptic-cylinder waveguides is also addressed. It is shown that phase speeds are shifted by an amount proportional to v_z ($-v_z$) for flow along (against) the direction of sound propagation independent of the sound propagation mode in consideration. Symmetry properties and nodal characteristics of pressure eigenmodes in terms of the Mathieu functions and the modified Mathieu functions are also discussed. The work presented is application-oriented and the application in hand has important implications for flow-acoustics devices.

1. Introduction

The Helmholtz equation is known to be separable in 11 coordinate systems including the elliptic cylinder coordinates (ECC) [1, 2, 3]. In this paper, the theory of acoustics in waveguides with an elliptical cross section is presented subject to rigid-wall boundary conditions (Neumann boundary conditions). The influence of a medium mean flow in an elliptic cylinder on acoustic propagation properties is also addressed. Furthermore, we briefly discuss eigenmodes in acoustical enclosures formed by a finite-length elliptical cylinder and two plane surfaces ($z = z_1$ and $z = z_2$) [4]. Similar studies of acoustic enclosures in parabolic cylinder [5] and parabolic rotational coordinates have been recently carried out [6]. Related studies of waveguides and cavity resonators (including elliptic-cylinder geometries) play an essential role in microwave physics [7, 8, 9, 10] as they are widely used in physical measurements and special devices. A recent examination of confined electronic states in flat elliptic semiconductor quantum dots and wires [11, 12] emphasizes the interest in understanding wave propagation phenomena in ECC. Although, in the case with mean flow, the governing differential equation is different from the Helmholtz equation, it is still possible to separate the problem. Some implications for ultrasonic flow measurement using the transit-time method are given. Finally, considerable attention is given to the study of symmetry properties and nodal characteristics of eigenmodes in ECC.

In the next section, we show how the acoustic wave equation in a medium with mean flow confined by elliptic-cylinder walls is separated using elliptic cylinder coordinates and give the

[1] corresponding author

solution of the resulting ordinary differential equations. We then present numerical results for various eigenmodes. It is necessary to emphasize that the work presented is application-oriented with important implications for ultrasonic flow measurement.

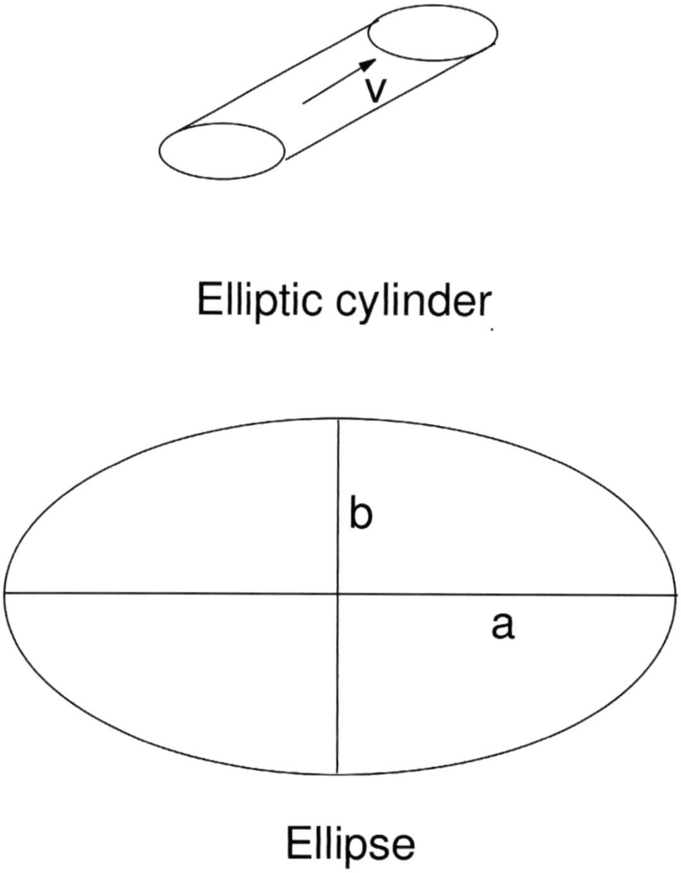

Figure 1. Elliptic cylinder and cross-sectional view (ellipse).

2. Theory

Consider in the following sound propagation in a moving fluid in a duct with an elliptical cross-section. Let us assume that the fluid speed $\vec{v}$ is constant in space and time and that the speed of sound c and fluid mass density ρ_0 are constants in space corresponding to the homogeneous case. In this case, the resulting (adiabatic) wave equation becomes [13, 14]

$$\vec{\nabla}^2\phi - \frac{1}{c^2}D_t^2\phi = 0, \tag{1}$$

where

$$p = -\rho_0 D_t\phi$$
$$D_t = \frac{\partial}{\partial t} + \vec{v}\cdot\vec{\nabla}. \tag{2}$$

Here, ϕ, p, and t are the velocity potential, the acoustic pressure, and time, respectively. Further simplifications are possible if monofrequency conditions are assumed and if the fluid flow is directed along the duct axis, i.e., $\vec{v} = (\vec{v}_{\parallel}, v_z) = (0, v_z)$, where $\vec{v}_{\parallel}$ is the mean-flow component in the plane defined by the duct cross section. Then, Equation (1) reduces to [for the spatial part of the potential field following a trivial division by the monofrequency term $\exp(i\omega t)$ in all factors]

$$\vec{\nabla}^2\phi + \frac{\omega^2}{c^2}\phi - \frac{2i\omega}{c^2}v_z\frac{\partial\phi}{\partial z} = 0, \tag{3}$$

where ω $(= 2\pi f)$ is the angular sound frequency and terms of second order in the velocity v_z have been discarded (a reasonable approximation since the flow velocity is small for typically liquid flows, i.e., less than 1 m/sec while the speed of sound is approximately three orders of magnitude higher higher). Equation (3) is conveniently written in elliptic-cylinder coordinates u, v, and z given the geometry at hand (Figure 1). Elliptic-cylinder coordinates are related to Cartesian coordinates x, y, and z as follows [15]:

$$\begin{aligned} x &= f\cosh u\cos v \\ y &= f\sinh u\sin v \\ z &= z \\ f &= \sqrt{a^2 - b^2} \\ 0 \le u \le \infty; 0 \le v &\le 2\pi; -\infty \le z \le \infty, \end{aligned} \tag{4}$$

with f the focal length and $2a$ $(2b)$ is the long (short) axis of the ellipse $(a > b)$. One immediately observes from Equation (4)that constant u curves form elliptic cylinders in Cartesian coordinates while constant v curves form hyperbolic cylinders in Cartesian coordinates. The boundary of the ellipse corresponds to $u = U = \text{arctanh}(b/a)$. Equation (3), written in elliptic-cylinder coordinates, becomes

$$\frac{2}{f^2(\cosh 2u - \cos 2v)}\left[\frac{\partial^2}{\partial u^2} + \frac{\partial^2}{\partial v^2}\right]\phi + \frac{\partial^2\phi}{\partial z^2} + \frac{\omega^2}{c^2}\phi - \frac{2i\omega}{c^2}v_z\frac{\partial\phi}{\partial z} = 0. \tag{5}$$

Proposing a separable solution

$$\phi(u, v, z) = U(u)V(v)Z(z), \tag{6}$$

Equation (5) can be rewritten as three ordinary differential equations

$$\frac{d^2V}{dv^2} + [\beta - 2\gamma\cos 2v]\,V = 0 \tag{7}$$

$$\frac{d^2U}{du^2} - [\beta - 2\gamma\cosh 2u]\,U = 0 \tag{8}$$

$$\frac{d^2Z}{dz^2} - \frac{2i\omega}{c^2}v_z\frac{dZ}{dz} + \alpha Z = 0 \tag{9}$$

where

$$\left(\frac{\omega^2}{c^2} - \alpha\right)\frac{f^2}{2} = 2\gamma, \tag{10}$$

and α, β are separation constants. The equation in Z can be solved immediately so as to give

$$Z(z) = Z'\exp(i\delta z), \tag{11}$$

with Z' a constant and (to first order in v_z)

$$\delta = \frac{\omega}{c}\frac{v_z}{c} \pm \sqrt{\alpha}. \tag{12}$$

Consider next the equation in V (the Mathieu equation). Physically acceptable solutions must be periodic of period π or 2π. A countably infinite sequence of characteristic values $\beta_l^1(\gamma)$ yield even periodic solutions $[ce_l(v,\gamma)]$. There is another countably infinite sequence of characteristic values $\beta_{l+1}^2(\gamma)$ which yield odd periodic solutions $[se_{l+1}(v,\gamma)]$. In the latter expressions, l is an integer ($l = 0, 1, 2, 3, ...$). For each γ and characteristic value [either $\beta_l^1(\gamma)$ or $\beta_{l+1}^2(\gamma)$], two solutions to the Mathieu Equation (7) exist because it is a second-order differential equation. One solution is periodic: $ce_l(v,\gamma)$ [or $se_{l+1}(v,\gamma)$] while the other $fe_l(v,\gamma)$ [or $ge_{l+1}(v,\gamma)$] is not a periodic function. Hence, solutions of the form $fe_l(v,\gamma)$ [or $ge_{l+1}(v,\gamma)$] cannot be solutions to the problem at hand and will be discarded in the following.

Consider now the equation in U (the modified Mathieu equation). The characteristic values $\beta_l^1(\gamma)$ [or $\beta_{l+1}^2(\gamma)$] and γ determined by solving the Mathieu equation also appear as coefficients in the modified equation. The complete solution of the modified Mathieu equation is

$$U(u) = A\ Ce_l(u,\gamma) + B\ Fe_l(u,\gamma), \tag{13}$$

corresponding to characteristic values $\beta_l^1(\gamma)$, or

$$U(u) = C\ Se_{l+1}(u,\gamma) + D\ Ge_{l+1}(u,\gamma), \tag{14}$$

corresponding to characteristic values $\beta_{l+1}^2(\gamma)$, where Ce_l, Fe_l, Se_{l+1}, and Ge_{l+1} are modified Mathieu functions and A, B, C, and D are constants. Again, one solution Fe_l [or Ge_{l+1}] can be discarded on physical reasoning. This is due to the fact that the velocity potential (or acoustic pressure) must be differentiable continuous across the focal line $u = 0$. Mathematically, this is stated as [3]

$$\lim_{v\to 0} \phi(u=0,v,z) = \lim_{v\to 0} \phi(u=0,-v,z), \tag{15}$$

and

$$\lim_{v\to 0} \frac{\partial\phi}{\partial u}|(u=0,v,z) = -\lim_{v\to 0} \frac{\partial\phi}{\partial u}|(u=0,-v,z). \tag{16}$$

These requirements are fulfilled only if the coefficients B and D are zero in Equations (13) and (14). In conclusion, acceptable solutions have the form

$$\phi_\gamma^e(u,v,z) = \sum_{l=0}^{\infty} \phi_{l,\gamma}^e(u,v,z) = \sum_{l=0}^{\infty} N_l ce_l(v,\gamma) Ce_l(u,\gamma) \exp(i\delta^\pm z), \tag{17}$$

(even solutions) or

$$\phi_\gamma^o(u,v,z) = \sum_{l=0}^{\infty} \phi_{l+1,\gamma}^o(u,v,z) = \sum_{l=0}^{\infty} N'_{l+1} se_{l+1}(v,\gamma) Se_{l+1}(u,\gamma) \exp(i\delta^\pm z), \tag{18}$$

(odd solutions) with

$$\gamma = \left(\frac{\omega^2}{c^2} - \alpha\right)\frac{f^2}{4} \tag{19}$$

$$\delta^\pm = \frac{\omega}{c}\frac{v_z}{c} \pm \sqrt{\alpha}. \tag{20}$$

Here, N_l, N'_{l+1} are constants. The set of possible γ constants remains to be found. This is done by imposing Neumann boundary conditions on the velocity potential ϕ along the boundary of the ellipse $u = U = \text{arctanh}(b/a)$ corresponding to the rigid-wall case, i.e.,

$$\frac{dU}{du}|(u = U) = 0, \tag{21}$$

or

$$\frac{dCe_l(u, \gamma^1_{i,l})}{du}|(u = U) = 0$$
$$\frac{dSe_{l+1}(u, \gamma^2_{i,l+1})}{du}|(u = U) = 0, \tag{22}$$

where γ values are ordered such that $\gamma^k_{i,l} < \gamma^k_{i+1,l}$ $(k = 1, 2 \text{ and } i = 1, 2, ...)$.

Thus, in the process of finding solutions, the following procedure is carried out in the waveguide case: Given ω, c, a, and b, Equation (22) is solved so as to find $\gamma^k_{i,l}$. Insertion into Equation (20) then gives $\alpha^k_{i,l}$ and $\delta^{k\pm}_{i,l}$. Note that the influence of a background flow v_z is simply to shift $\delta^{k+}_{i,l}$ ($\delta^{k-}_{i,l}$) by an amount $\frac{\omega}{c}\frac{v_z}{c}$ ($-\frac{\omega}{c}\frac{v_z}{c}$). The two sign possibilities reflect the fact that sound may propagate along (against) the direction of flow. All other parameters are left unaffected by the presence of a mean flow. Let us define the phase difference $\Delta\psi^k_{i,l}$ between an upstream and a downstream sound propagation case as [17]:

$$\Delta\psi^k_{i,l} = \omega b_a \left(\frac{1}{c^{k-}_{pi,l}} - \frac{1}{c^{k+}_{pi,l}} \right), \tag{23}$$

where b_a denotes the distance between the two transducers, and $c^{k-}_{pi,l} = \frac{\omega}{\sqrt{\alpha^k_{i,l} + \frac{\omega}{c}\frac{v_z}{c}}}$ $\left(c^{k+}_{pi,l} = \frac{\omega}{\sqrt{\alpha^k_{i,l} - \frac{\omega}{c}\frac{v_z}{c}}} \right)$ are the phase speeds corresponding to an upstream (downstream) sound propagation case, respectively. Equation (23) is conveniently rewritten as

$$\Delta\psi^k_{i,l} = 2b_a \frac{\omega}{c}\frac{v_z}{c}. \tag{24}$$

The transit time difference $\Delta t^k_{i,l}$ between the two successive sound propagation situations then becomes:

$$\Delta t^k_{i,l} = \frac{\Delta\psi^k_{i,l}}{\omega}, \tag{25}$$

and the so-called deviation of measurement $E^k_{i,l}$, often used as a quality measure for an ultrasonic flow meter, can be written as [17]:

$$E^k_{i,l} = \frac{\Delta t^k_{i,l} - 2b_a\bar{v}_z/c^2}{2b_a\bar{v}_z/c^2}, \tag{26}$$

where $2b_a\bar{v}_z/c^2$ is the transit time difference of the fundamental mode. It is evident, that the right-handside of Equation (26) is zero for all modes by employing Equations (24) and (25). Hence, ultrasonic flow measurement is, theoretically, error-free for sound propagation in elliptical waveguides in the homogeneous flow case. In real flow situations, the flow velocity depends on the in-plane position coordinate, and in this case, it is unlikely that ultrasonic flow measurement is characterized by a deviation of measurement which is zero (bearing in mind the corresponding conclusions of flow measurement in cylindrical waveguides [17, 18]).

2.1. Acoustic eigenmodes in elliptic-cylinder enclosures

Consider next an acoustic enclosure consisting of a finite-length elliptic cylinder (with length L) and two plane $z = const$ surfaces. In this case, there is no mean flow, i.e., $v_z = 0$. The z axis is chosen such that the two plane surfaces correspond to $z = 0$ and $z = L$, respectively.

The general solution for Z is, refer to Equations (17) and (18),

$$Z(z) = E\exp(i\delta^+ z) + F\exp(i\delta^- z), \tag{27}$$

where E and F are constants. Imposing the boundary conditions:

$$\frac{dZ}{dz}|(z=0) = \frac{dZ}{dz}|(z=L) = 0, \tag{28}$$

leads to

$$\begin{aligned} \alpha &= \left(\frac{n\pi}{L}\right)^2, \\ \delta^+ &= -\delta^- = \sqrt{\alpha}, \end{aligned} \tag{29}$$

with n an integer. Insertion of Equation (29) in the first of Equations (20) allows the eigenfrequencies f^k_{iln} to be determined

$$f^k_{iln} = \frac{c}{2\pi}\sqrt{\frac{4}{f^2}\gamma^k_{i,l} + \left(\frac{n\pi}{L}\right)^2}. \tag{30}$$

Thus eigenfrequencies of an elliptic-cylinder acoustic enclosure depend on the indices i, l, k, and n as expected. The $\gamma^k_{i,l}$ values are the same as in the elliptic-cylinder waveguide case since $\gamma^k_{i,l}$ is a function of the geometry of the ellipse being common to the waveguide and enclosure.

Note, finally, that the wave equation in the presence of flow [Equation (3)] can be transformed into the Helmholtz equation using a Prandtl-Glauert-like transformation:

$$\begin{aligned} x' &= x, \\ y' &= y, \\ z' &= \frac{z}{1 - v_z^2/c^2}, \\ \phi(x,y,z) = \phi'(x',y',z')\exp&\left[\frac{i\omega z' v_z}{c\sqrt{c^2 - v_z^2}}\right]. \end{aligned} \tag{31}$$

Hence, if the Helmholtz equation is separable in a coordinate system then so is the wave equation with mean flow.

3. Results and Discussions

In this Section, calculated results of acoustic eigenmodes in elliptical ducts of various b/a ratios are described. The fluid in the duct is assumed to be water. We employ the following parameter values corresponding to the excitation of ultrasonic waves in water: $f = 1 \times 10^6$ Hz, $a = 1$ cm, $c = 1500$ m/sec, and b values in the range: $2 - 9.9$ mm. Mathieu functions and modified Mathieu functions and associated characteristic values are computed using the FORTRAN routines in Ref. [16]. In Figure 2a, we show the computed values of the smallest value of $\gamma^1_{1,l}$ for a given l: $\gamma^1_{1,l}$ for $l = 0, 1, 2, 3$, and 4 (even solutions) as a function of the b/a ratio. The asterisk, circle, diamond, square, and triangle data corresponds to $l = 0, 1, 2, 3$, and 4, respectively. It is evident that, when $l \geq 1$, the smallest number of γ^1_l increases with the value of l for all b/a ratios.

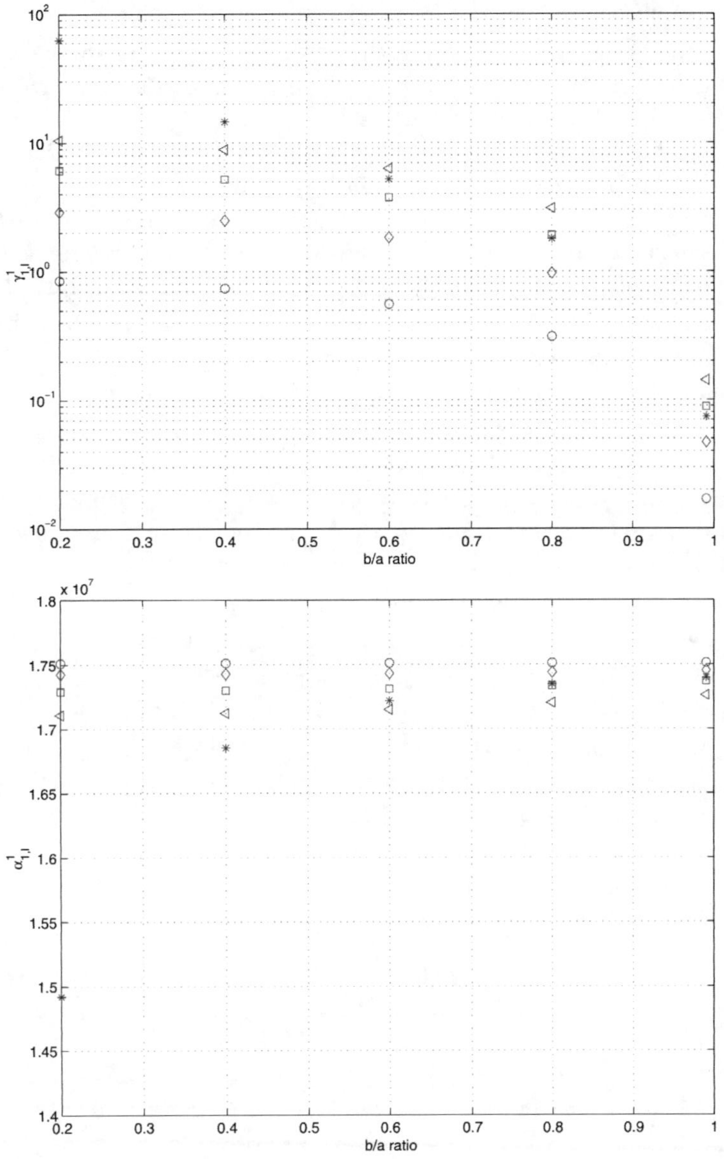

Figure 2. Plot of $\gamma^1_{1,l}$ [upper figure (a)] and $\alpha^1_{1,l}$ [lower figure (b)] as a function of the b/a ratio for $l = 0, 1, 2, 3$ and 4 for the case with even modes. The asterisk, circle, diamond, square, and triangle data correspond to $l = 0, 1, 2, 3$, and 4, respectively.

In actual fact, the number of zeros in the interval $0 < v < \pi$ of $ce_l(v, \gamma_{1,l}^1)$ and $se_l(v, \gamma_{1,l+1}^2)$ is exactly l [19]. Thus, the increase of $\gamma_{1,l}^1$ with increasing l reflects the case well-known from spherical and cylindrical geometries that the in-plane (cross-sectional) wave number increases with the number of nodes of the angular part of the eigenfunction in the range $0 < v < \pi$. We also notice that $\gamma_{1,l}^1$ for $l = 0$ is higher than $\gamma_{1,l}^1$ value for $l = 1, 2, 3, 4$ when $b/a < 0.4$. This result is a consequence of $\phi_{l=0}^e(u, v, z)$ [in the following, we omit the subscript γ on $\phi_l^{e,o}(u, v, z)$ to simplify the notation] being the second (first non-trivial) eigenfunction for $l = 0$ and b being considerably smaller than a. The trivial solution is constant and corresponds to $\gamma = 0$. Further details follow in the discussion of Figure 4 later in this Section. In Figure 2b, $\alpha_{1,l}^1$ values are plotted as a function of the b/a ratio corresponding to $\gamma_{1,l}^1$ values depicted in Figure 2a. The values of $\alpha_{1,l}^1$ are computed based on Equation (20).

In Figure 3a, we show the smallest value of $\gamma_{1,l+1}^2$ for $l = 0, 1, 2, 3$, and 4 as a function of the b/a ratio. As before, the asterisk, circle, diamond, square, and triangle data correspond to $l = 0, 1, 2, 3$, and 4, respectively. It is evident that $\gamma_{1,l+1}^2$ increases with increasing l for all b/a ratios in the range $0.2 \leq b/a \leq 0.99$. In contrast to the case with even Mathieu functions and modified Mathieu functions, the present case with odd Mathieu functions and modified Mathieu functions is characterized by the following: The eigenvalue $\gamma_{1,l+1}^2$ for each l corresponds to an eigenfunction $(se_{l+1}(v, \gamma_{1,l+1}^2)$ having l nodes in the range $0 < v < \pi$ while $Se_{l+1}(u, \gamma_{1,l+1}^2)$ has no nodes in the range $0 \leq u \leq \mathrm{arctanh}(b/a)$. This explains why $\gamma_{1,l+1}^2$ increases with l for $l = 0, 1, 2, 3$, and 4. In Figure 3b, $\alpha_{1,l+1}^2$ values are plotted against the b/a ratio corresponding to the $\gamma_{1,l+1}^2$ values shown in Figure 3a. The values of $\alpha_{1,l+1}^2$ are computed based on Equation (20).

In Figure 4, a cross-sectional plot of the first even eigenmode solutions $[ce_l(v, \gamma_{1,l}^1)Ce(u, \gamma_{1,l}^1)]$ for each l in the range $l = 0, 1, 2, 3$, and 4 is shown. The b/a ratio is 0.2 and the major axis is along x. Let us first discuss symmetry properties. The even eigenmodes ϕ_l^e obey the following symmetry relation

$$\phi_l^e(x, y, z) = \phi_l^e(u, v, z) = \phi_l^e(u, -v, z) = \phi_l^e(x, -y, z), \tag{32}$$

because $ce_l(-v, \gamma_l^1) = ce_l(v, \gamma_l^1)$, i.e., even eigenmodes ϕ_l^e are symmetric with respect to a mirror reflection in the $y = 0$ plane. Similarly, in the case of ϕ_{2l}^e ($l = 0, 1, ...$), the following relation applies

$$\phi_{2l}^e(x, y, z) = \phi_{2l}^e(u, v, z) = \phi_{2l}^e(u, \pi - v, z) = \phi_{2l}^e(-x, y, z), \tag{33}$$

because $ce_{2l}(\pi - v, \gamma_{1,2l}^1) = ce_{2l}(v, \gamma_{1,2l}^1)$, i.e., $\phi_{2l}^e(x, y, z)$ ($l = 0, 1, ...$) is symmetric with respect to a mirror reflection in the $x = 0$ plane. In the case of $\phi_{2l+1}^e(x, y, z)$ ($l = 0, 1, ...$), one finds

$$\phi_{2l+1}^e(x, y, z) = \phi_{2l+1}^e(u, v, z) = -\phi_{2l+1}^e(u, \pi - v, z) = -\phi_{2l+1}^e(-x, y, z), \tag{34}$$

because $ce_{2l+1}(\pi - v, \gamma_{1,2l+1}^1) = -ce_{2l+1}(v, \gamma_{1,2l+1}^1)$, i.e., $\phi_{2l+1}^e(x, y, z)$ ($l = 0, 1, ...$) is antisymmetric with respect to a mirror reflection in the $x = 0$ plane.

Consider next the odd eigenmodes ϕ_l^o. Due to $se_l(v, \gamma_{l+1}^2)$ being an odd function, it is found that

$$\phi_l^o(x, y, z) = \phi_l^o(u, v, z) = -\phi_l^o(u, -v, z) = -\phi_l^o(x, -y, z), \tag{35}$$

i.e., ϕ_l^o are antisymmetric with respect to a mirror reflection in the $y = 0$ plane. Similarly,

$$\phi_{2l}^o(x, y, z) = \phi_{2l}^o(u, v, z) = -\phi_{2l}^o(u, \pi - v, z) = -\phi_{2l}^o(-x, y, z), \tag{36}$$

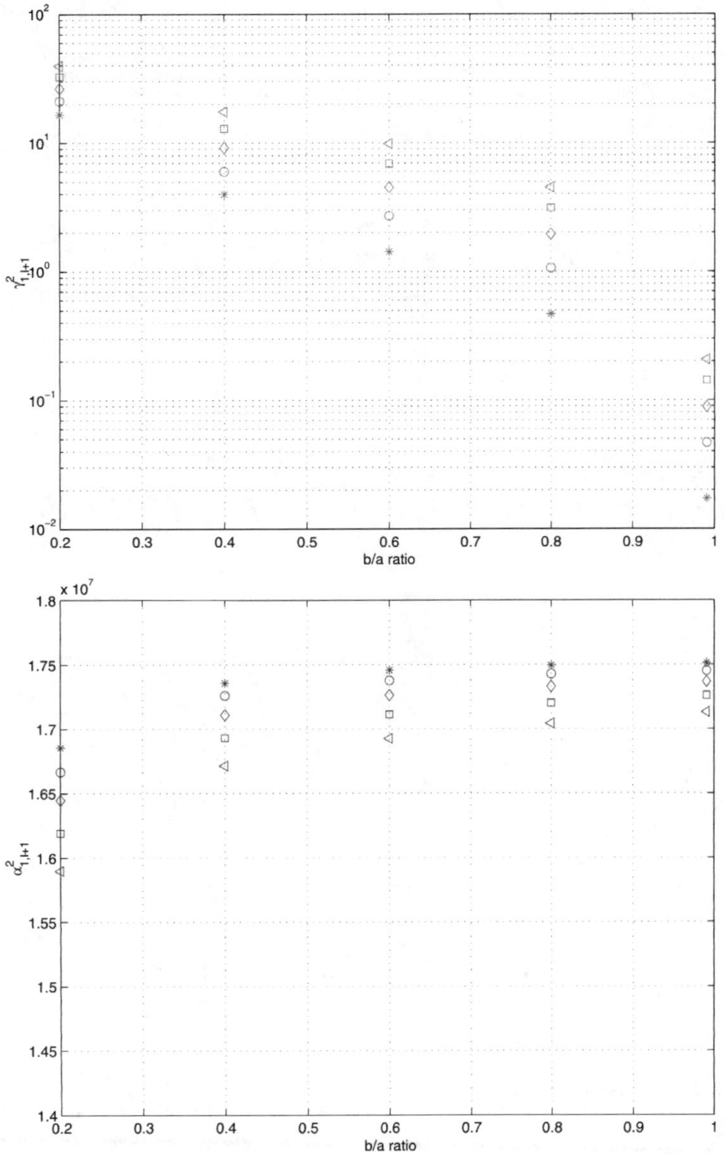

Figure 3. Plot of $\gamma^2_{1,l+1}$ [upper figure (a)] and $\alpha^2_{1,l+1}$ [lower figure (b)] as a function of the b/a ratio for $l = 0, 1, 2, 3$ and 4 for the case with odd modes. The asterisk, circle, diamond, square, and triangle data correspond to $l = 0, 1, 2, 3,$ and 4, respectively.

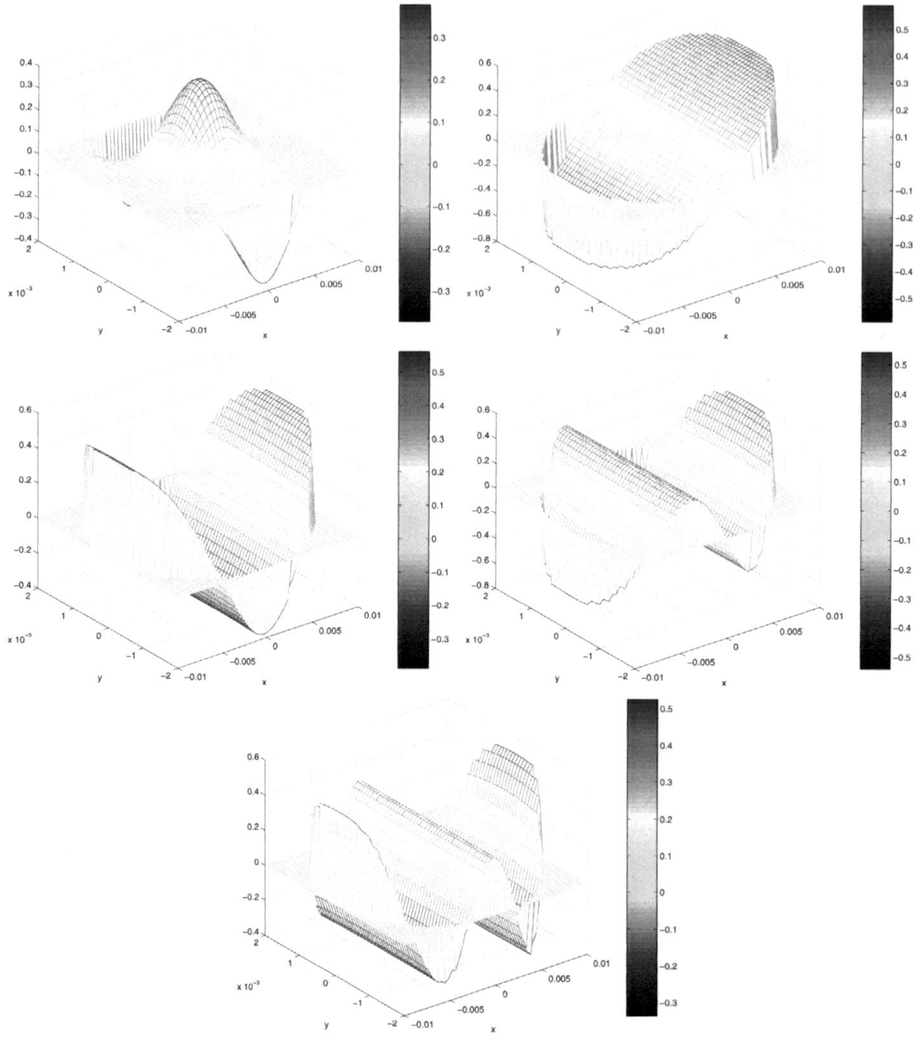

Figure 4. The first five even eigenmodes corresponding to $\gamma^1_{1,l}$ plotted as a function of x and y (ellipse). The upper-left, upper-right, middle-left, middle-right, and lower figures correspond to $l = 0, 1, 2, 3,$ and 4, respectively.

and

$$\phi^o_{2l+1}(x, y, z) = \phi^o_{2l+1}(u, v, z) = \phi^o_{2l+1}(u, \pi - v, z) = \phi^o_{2l+1}(-x, y, z), \tag{37}$$

i.e., ϕ^o_{2l} (ϕ^o_{2l+1}) is antisymmetric (symmetric) with respect to a mirror reflection in the $x = 0$ plane ($l = 0, 1, 2, ...$).

The above-mentioned symmetry properties can be observed in Figure 4 for the even eigenmodes. In Figure 4, the upper-left, upper-right, middle-left, middle-right, and lower figures are for $l = 0, 1, 2, 3,$ and 4, respectively. As mentioned earlier, the number of nodes of ϕ^e_l in the interval $0 < v < \pi$ is exactly l. Hence, l nodal lines exist in the $x - y$ plane for ϕ^e_l and appear as parabolic sheets (because $v = constant \Rightarrow \left[\frac{x}{\cos v}\right]^2 - \left[\frac{y}{\sin v}\right]^2 = f^2 \Rightarrow$ parabolic sheets). This is clearly observed in Figure 4. Note, in particular, that the upper-left figure for $l = 0$ is zero

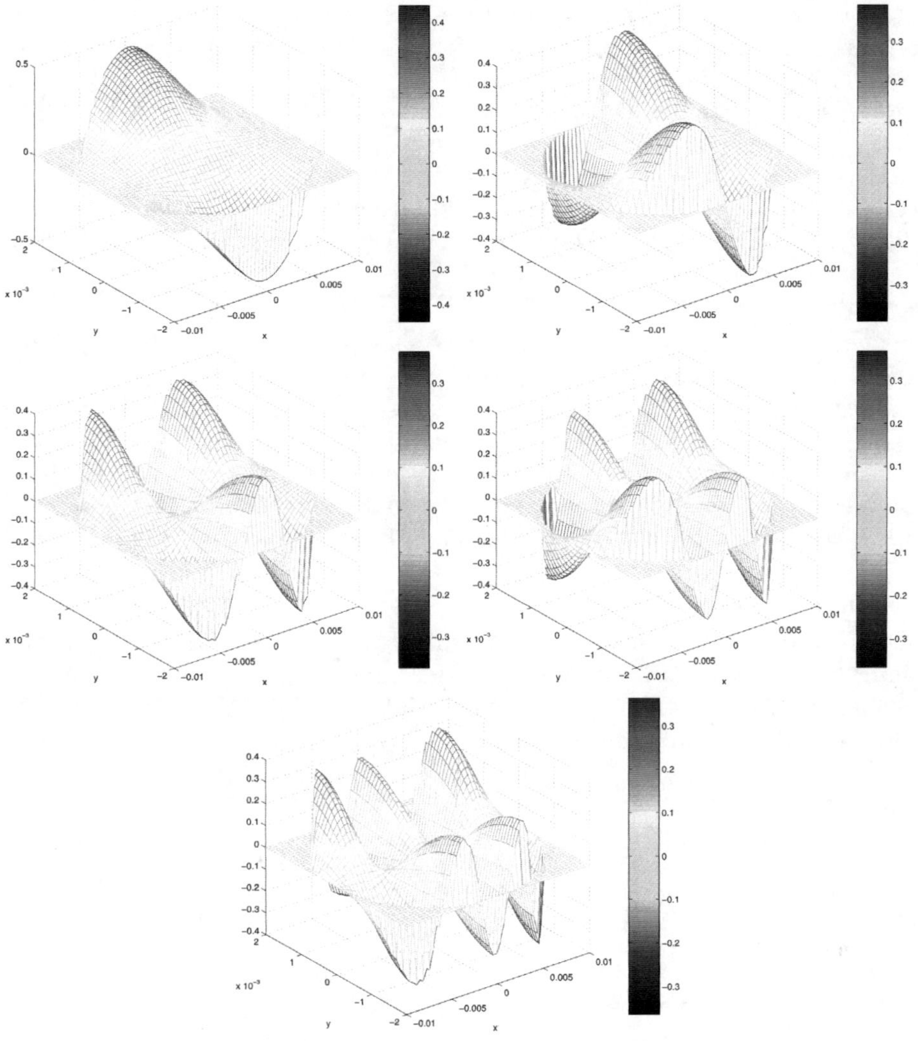

Figure 5. The first five odd eigenmodes corresponding to $\gamma^2_{1,l+1}$ plotted as a function of x and y (ellipse). The upper-left, upper-right, middle-left, middle-right, and lower figures correspond to $l = 0, 1, 2, 3,$ and 4, respectively.

along an elliptic sheet where $u = constant$ (because $u = constant \Rightarrow \left[\frac{x}{\cosh u}\right]^2 + \left[\frac{y}{\sinh u}\right]^2 = f^2 \Rightarrow$ elliptic sheets). The reason is that the solution for $l = 0$ is the second solution for which $\frac{dCe_0}{du}|(u = U) = 0$, and this solution satisfies $Ce_0(u') = 0$ for a value $u = u' < U$ (keep in mind that the first solution satisfying $\frac{dCe_0}{du}|(u = U) = 0$ is a constant function).

In Figure 5, a cross-sectional plot of the first odd eigenmode $se_{l+1}(v, \gamma^2_{1,l+1})Se_{l+1}(u, \gamma^2_{1,l+1})$ for each l ($l = 0, 1, 2, 3,$ and 4) is shown corresponding to a b/a ratio equal to 0.2. It is clearly observable that all functions are odd in v (or y) and that the number of zeroes for $se_{l+1}(v, \gamma^2_{1,l+1})$ is l in the range $0 < v < 2\pi$.

Finally, we briefly compare our solutions to those of Hong and Kim [4]. Specifically, we compare our modes in Figures 4 and 5 to their Figure 2. Note that the level ordering need not be the same if the eccentricities differ due to the phenomenon of level crossing [12]. Nevertheless, we find that, apart from one mode, we can relate our modes to theirs with the same ordering. For even modes, our l index corresponds to their m; for odd states, our l index corresponds to $m-1$. Thus, our modes in Figure 4 (taking them from left to right and from top to bottom, except for the lowest mode) correspond to the following of their modes: $P(1,1)1$, $P(2,1)1$, $P(3,1)1$, $P(4,1)1$. The one difference lies in the lowest mode where they give the trivial solution; we ignore the trivial solution. In addition, we have found another mode that is absent in their case. Similarly, our modes in Figure 5 (taking them from left to right and from top to bottom) correspond to the following of their modes: $P(1,1)2$, $P(2,1)2$, $P(3,1)2$, $P(4,1)2$, $P(5,1)2$.

4. Conclusion

Sound propagation in elliptic-cylinder waveguides and acoustic enclosures is presented. The influence of a medium mean flow on acoustic pressure is included in the general theory and implications for ultrasonic flow measurement based on the transit-time method is briefly mentioned. It is shown that phase speeds are shifted by an amount proportional to v_z $(-v_z)$ for flow along (against) the direction of sound propagation independent of the sound propagation mode in consideration. Symmetry properties and nodal characteristics of pressure eigenmodes in terms of the Mathieu functions and the modified Mathieu functions are also discussed.

5. Acknowledgements

The authors wish to express their gratitude to one of the reviewers for an insightful comment on the Prandtl-Glauert-like transformation transforming the wave equation under flow to the Helmholtz equation. Work of LCLYV was supported by an NSF CAREER award (NSF Grant No. 9984059) and a Balslev award (Denmark).

6. References

[1] P. M. Morse and H. Feshbach, *Methods of Theoretical Physics* (McGraw-Hill, New York, 1953).
[2] P. Moon and D. E. Spencer, *Field Theory Handbook* (Springer-Verlag, Berlin, 1961).
[3] N. W. McLachlan, *Theory and Application of Mathieu Functions* (Oxford University Press, London, 1947).
[4] K. Hong and J. Kim, "Natural mode analysis of hollow and annular elliptical cylinder cavities," J. Sound Vib. **183**, 327-351 (1995).
[5] M. Willatzen and L.C. Lew Yan Voon, "Acoustic eigenmodes in parabolic cylindrical structures," Proceedings of the Tenth International Congress on Sound and Vibration, July 7-10, Stockholm, Sweden, 3279-3286 (2003).
[6] M. Willatzen and L.C. Lew Yan Voon, "Quantum dot and acoustic enclosure problems in lens-shaped structures," Computational Science and Its Applications, LNCS 2668, Part II, 735-744 (2003).
[7] F. E. Borgnis and C. H. Papas, *Electromagnetic Waveguides and Resonators in Handbuch der Physik* (Springer-Verlag, Berlin, 1958).
[8] R. B. Dyott, *Elliptical Fiber Waveguides* (Artech House, Norwood MA, 1995).
[9] M. Schneider and J. Marquardt, "Fast computation of modified Mathieu functions applied to elliptical waveguide problems," IEEE Trans. on Microwave Theory and Techniques **47**, 513-516 (1999).
[10] S.-J. Zhang and Y.-C. Shen, "Eigenmode sequence for an elliptical waveguide with arbitrarily ellipticity," IEEE Trans. on Microwave Theory and Techniques **43**, 227-230 (1995).
[11] M. van den Broek and F. M. Peeters, "Confined states in two-dimensional flat elliptic quantum dots and elliptic quantum wires," Physica E **11**, 345-355 (2001).
[12] L. C. Lew Yan Voon, C. Galeriu, and M. Willatzen, "Comment on "Confined states in two-dimensional flat elliptic quantum dots and elliptic quantum wires," Physica E **18**, 547-549 (2003).
[13] D. I. Blokhintzev, *Acoustics of a Non-Homogeneous Moving Medium* (Gostekhizdat, Moscow, 1946).
[14] A. D. Pierce, "Wave equation for sound in fluids with unsteady inhomogeneous flow," J. Acoust. Soc. Am **87**, 2292-2299 (1990).
[15] M. Abramowitz and I. Stegun, *Handbook of Mathematical Functions* (Dover, New York, 1968).
[16] S. Zhang and J. Jin, *Computation of Special Functions* (John Wiley and Sons, Inc., New York, 1996).
[17] M. Willatzen, "Perturbation theory applied to sound propagation in a flowing media confined by a cylindrical waveguide," J. Acoust. Soc. Am **109**, 102-107 (2001).

[18] M. Willatzen, "Sound propagation in a moving fluid confined by cylindrical walls - a comparison between an exact analysis and the local-plane-wave approximation," J. Sound Vibrat. **247**, 719-729 (2002).

[19] J. Meixner and F. W. Schäfke, *Mathieu-Funktionen und Sphäroid-funktionen* (Springer-Verlag, Berlin, 1954).

Institute of Physics Publishing
doi:10.1088/1742-6596/52/1/002

Journal of Physics: Conference Series **52** (2006) 14–26
Mathematics for Industry in Denmark

Detection of time of arrival of ultrasonic pulses

Carl D. Latino Ph.D.[1]
Associate Professor ECEN
Oklahoma State University
Stillwater, OK 74078, USA

Niels Lervad Andersen [2]
Associate Professor
University of Southern Denmark
Grundtvigs Allé 150
DK 6400 Soenderborg, Denmark

Frands Voss [3]
Director, MCI
University of Southern Denmark
Grundtvigs Allé 150
DK 6400 Soenderborg, Denmark

[1] Latino@okstate.edu, [2] NLA@ingsdb.sdu.dk [3] Voss@mci.sdu.dk

Abstract: Applications exist that employ transit-time measurements to determine distances to scattering objects. These include blood-flow measurements, fluid-level detection, non-destructive testing of materials, image processing of geometrical bodies etc. Many of these require extreme accuracy in measuring transit time through a medium. The system typically consists of a transmitter, a medium and a receiver. The signal sent must operate within the limitations imposed by this overall system. Under ideal conditions a finite duration pulse consisting of a sinusoidal signal is sent and received. If the transmitted signal contains an integer number of complete cycles of a single frequency, performing a correlation between transmitted and received signals can accurately recover the transit time information. The method has difficulties especially in the presence of noise. The problem is compounded if the desired accuracy is a small fraction of a single sinusoid period. Noise could cause the correlation results to be off by a complete cycle. For a one MHz signal, one cycle error corresponds to a one microsecond error. For applications requiring nanosecond accuracy, this is not acceptable. Varying the frequency content of the signal reduces the error but requires that the transmitter, receiver and medium all have a broader band spectrum. The improvements we propose here work within a narrow band while enhancing the chance of recovering the signal accurately.

Keywords: time of arrival, ultrasonic pulse, correlation, frequency sweep, Piezoelectrics, Signal Coding, Arrival Time Detection, Vibroseis.
PACS: 43.60 Acoustic Signal Processing, 43.85 Acoustical Measurements and Instrumentation, 43.88 Transduction, Acoustical Devices for the Generation and Reproduction of Sound.
Mathematics Subject Classification: 60Gxx:60G35 Applications (signal detection, filtering,etc.); 91Axx:91A28 Signaling, communication; 94Axx:94A12 Signal theory (characterization, reconstruction, etc.).

1. Introduction

When transmitting a pulse through a medium, the received signal is attenuated, time shifted and distorted. We define a pulse to be any signal of finite amplitude and temporal duration, composed of one or a combination of frequencies. Even if the pulse is designed to be of a single frequency, the mere act of turning a signal on and then off, introduces many unwanted frequencies. These frequencies make the problem of receiving the signal more complex but not intractable. Maximizing the energy transmitted in the form of higher amplitude or longer pulse duration improves the signal to noise ratio making signal reception more manageable. However, the system parameters and time constraints limit the pulse duration and maximum energy possible. Receiving a signal imbedded in noise is a task that can be handled effectively by correlation. Correlation is a commonly used mathematical operation where two signals are multiplied together and the result integrated over all time. If the signals are time shifted relative to each other the integration results in a function of time shift. The function has a maximum point corresponding to the time shift of "best alignment". For a given amount of energy, performing a correlation between transmitted and received pulse (cross correlation) gives an accurate measure of the delay time between the signals. The correlation plot of the transmitted and received pulses generally has multiple peaks, but the maximum absolute value occurs when the pulses align perfectly. For correlation we used C=XCORR(A,B), which is one of the many Matlab® functions. This function takes two vectors A and B, each of length N and returns a vector, C of length 2*N-1. Where C is a cross correlation sequence. In the case of A=B then the result is called the autocorrelation. This is the function we used to get many of our plots.

If the signal is transmitted without distortion, doing a cross correlation between the transmitted and received signals gives an accurate means of measuring time delay. This is accomplished by locating the absolute highest point of the cross correlation function. The time shift required to accomplish this maximum represents the transit time. Unfortunately, in the presence of noise and with bandwidth limitations of the system, the highest peak in the cross correlation may be difficult to distinguish from its nearest neighbors. It is desirable to increase the difference between the maximum peak and the secondary peaks in order to more reliably determine the true maximum peak and thus the true time delay. Working within the fixed limitations of the system, we propose a method, of improving the discrimination between the correlation peaks to improve the chance of identifying the true maximum.

An application requiring very accurate measurement is the measure of speed of sound through a liquid. Typical velocities are in the order of Mach 1. This method may be used to study the effects of pressure and other parameters on the speed of sound through the liquid in question. It may also be used to measure flow rates of the liquid. The restrictions imposed by the system are critical when it is necessary to measure transit times accurate in the nanosecond range. For example, if the driving frequency of the sinusoidal signal is 1 MHz, one period corresponds to one microsecond. If the speed in question is of the order of 300 meters per second, a nanosecond represents about 3 millimeters. If the desired accuracy is in the nanosecond range it must be possible to measure accurately within a small portion of a single cycle of the 1 MHz signal. This is approximately three orders of magnitude smaller time than a single period of the transmitted frequency. By appropriate selection of frequency and pulse duration, a cross correlation of the transmitted and received signals can very accurately recover the transit time. In practical applications the medium introduces not only a time delay, but also a distortion. The transmitter and receiver also have characteristics, which operate as filters, further introducing noise. The transmitter and receiver also often resonate at these frequencies. All these conditions complicate the problem and limit the effectiveness of the cross correlation.

2. System Limitations

Assuming the system is fixed and cannot be modified, improvement of the correlation can be accomplished by carefully crafting the transmitted pulse. A coded pulse technique of this type has been used by the oil exploration industry since the 1960's and is called Vibroseis [1]. Vibroseis uses large trucks to shake the ground with swept frequency signals and detect the reflected signals with seismometers. The choice of frequency sweep was dependent on many factors. A typical sweep duration was in the order of five to twenty seconds with frequencies from 5 to 100 Hz. Their application was therefore less restrictive than the one we are addressing. Unlike the Vibroseis[1] application, the system we are addressing uses low energy and narrow band transducers, which actually resonate at their natural frequency. If the application refers to measuring velocity through a medium, this velocity is affected by various parameters including flow direction and rate. For example, in flow measuring applications, the transducer is used for both receiving and transmitting the signal. This means that the pulse must travel with and against the flow and have a relatively short duration. A short pulse limits the amount of energy injected in the system and thus making it more susceptible to noise. Working within these constraints, we propose some solutions.

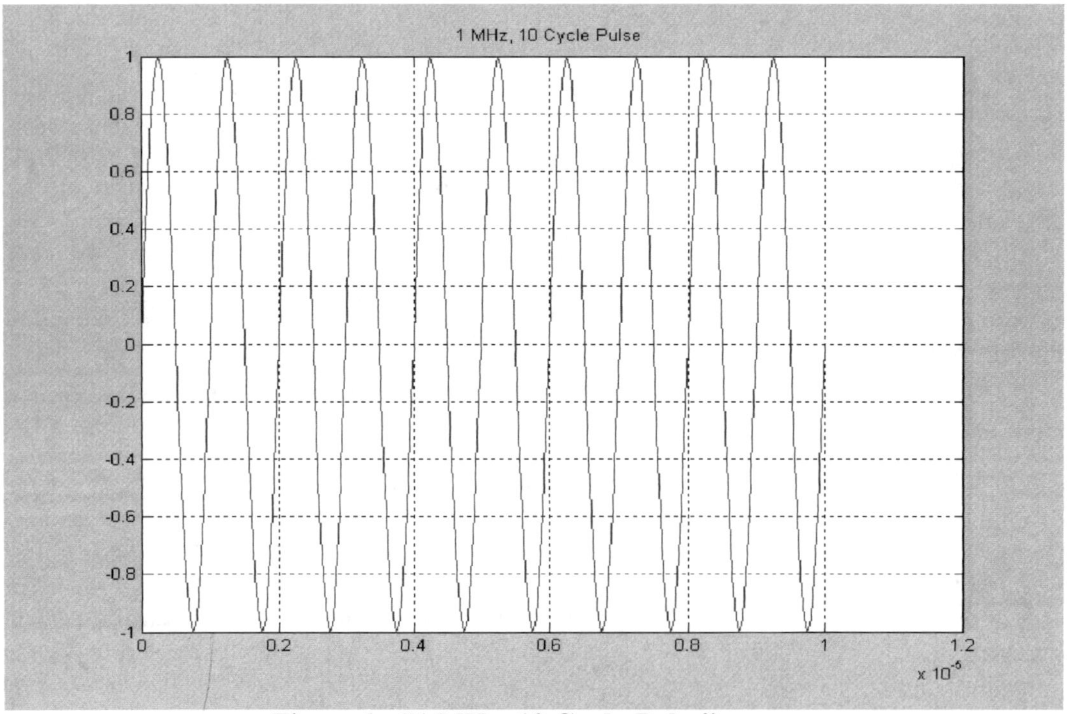

Figure 1. "1 MHz, 10 Cycle Pulse"

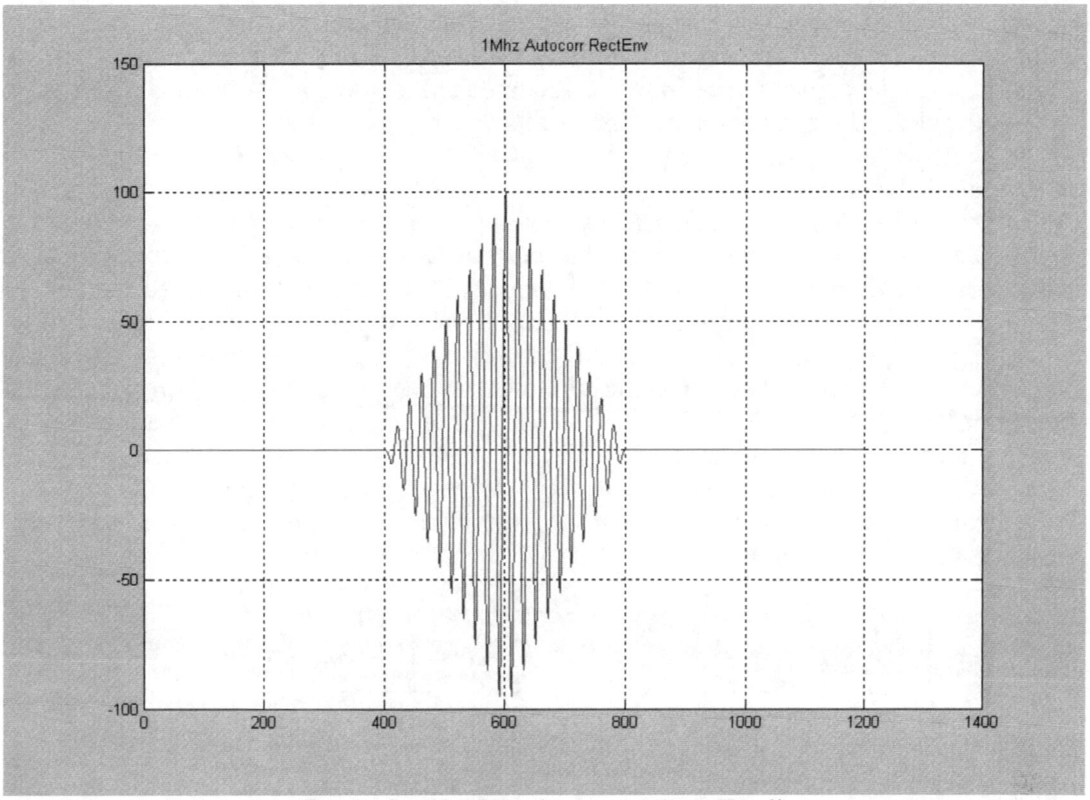

Figure 2. "1 MHz Autocorr RectEnv"

3. Limitations of Single Frequency Pulses

A single frequency sinusoid of finite duration has an autocorrelation whose value is a function of the number of complete cycles, N. We used Matlab® to do all our models and simulations for this paper. A single frequency example is illustrated in Figure 1. This figure illustrates a 1 MHz pulse of 10 microsecond duration. The value of the signal outside this range is zero. This particular 10 microsecond pulse was modeled using 200 sample points. The Autocorrelation is shown in Figure 2. The x axis corresponds to shift in sample points. Each sample point therefore corresponds to 50 nanoseconds. Thus, the 600 point on the x axis corresponds to a zero time shift, and maximum overlap, whereas 400 or less, or 800 or more are the points where the transmitted and received pulses have no overlap. The larger we make N, the larger the maximum correlation value. Making N too large, however, introduces the unwanted result that the secondary peaks of the autocorrelation are close in magnitude to the maximum. For example, if the maximum peak of the autocorrelation is K*N, the next peak has a magnitude K*(N-1). For large values of N, the two adjacent peaks are close in magnitude relatively measured and it is difficult to tell which one is the true maximum. Another problem with large N is that for a given frequency, it makes for a long pulse. The applications often limit the pulse duration. If N is too small, however, there may be insufficient energy that makes it to the receiver. Selecting N in the range of 10 to 20 and the frequency of about 1 MHz may be a good compromise. We will work under this assumption and use these numbers in our paper. Actual numbers selected must conform to the actual system parameters. With these numbers in mind there are other practical problems to be addressed. If the transducers used in the system are very narrow band they may resonate and oscillate even when no longer energized. To reduce the introduction of unwanted frequencies the pulse should be amplitude modulated with a suitable envelope. Although the envelope reduces the amount of total energy, it is advantageous because it reduces the amount of unwanted frequency components being transmitted. Quantitative analysis of the effects of envelopes, or "windows" on the signal can be found in fundamental textbooks on signal analysis such as [5], [6], [7], and are beyond the scope of this paper.

4. Proposed Pulse Designs

An improvement to the single frequency method is the use of a swept frequency pulse that operates within the bandwidth of the transducers in question. A swept frequency has a very unique shape and is easier to recognize at the receiver than a fixed frequency. Figure 3 illustrates such a pulse. The sudden start and stop of a frequency pulse, due to a rectangular envelope, has the additional disadvantage of generating unwanted frequencies. To reduce these unwanted frequencies the pulse must not only consist of the desired frequencies but also be combined with an appropriate envelope in order to transmit a more suitably shaped pulse. The envelope causes the overall pulse amplitude to smoothly grow from zero to peak value and descend smoothly at the end of the pulse. An unwanted result of the inclusion of the amplitude modulating envelope is that the total amount of energy transmitted is reduced.

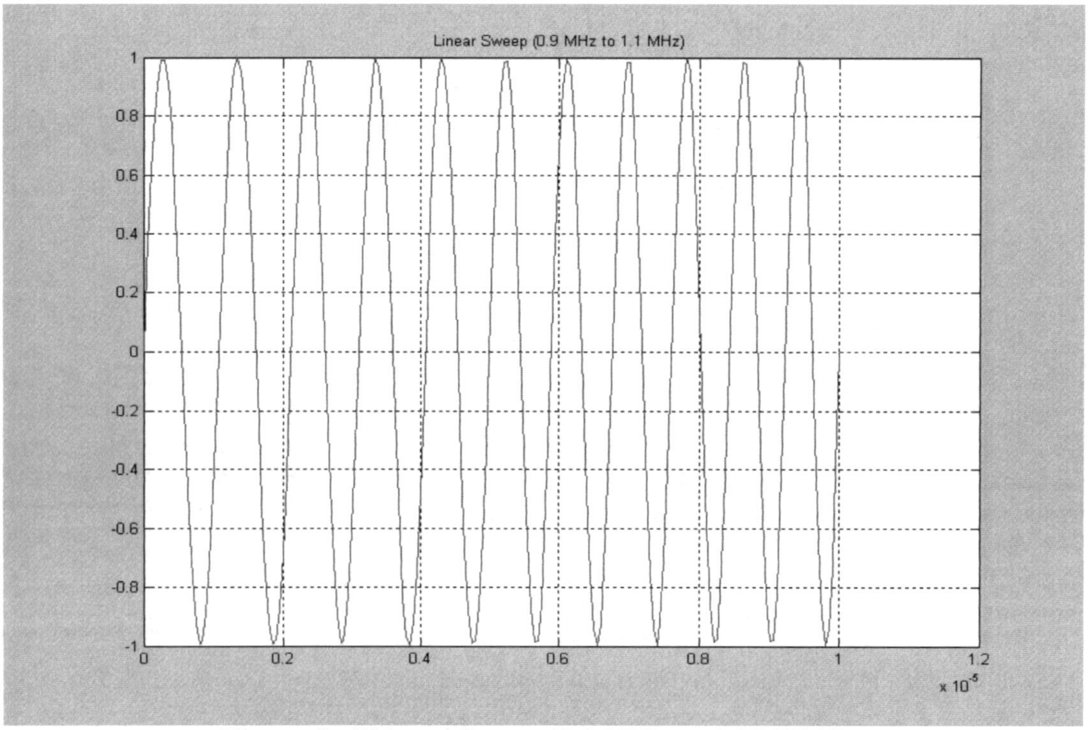

Figure 3. "Linear Sweep (0.9 MHz to 1.1 MHz)"

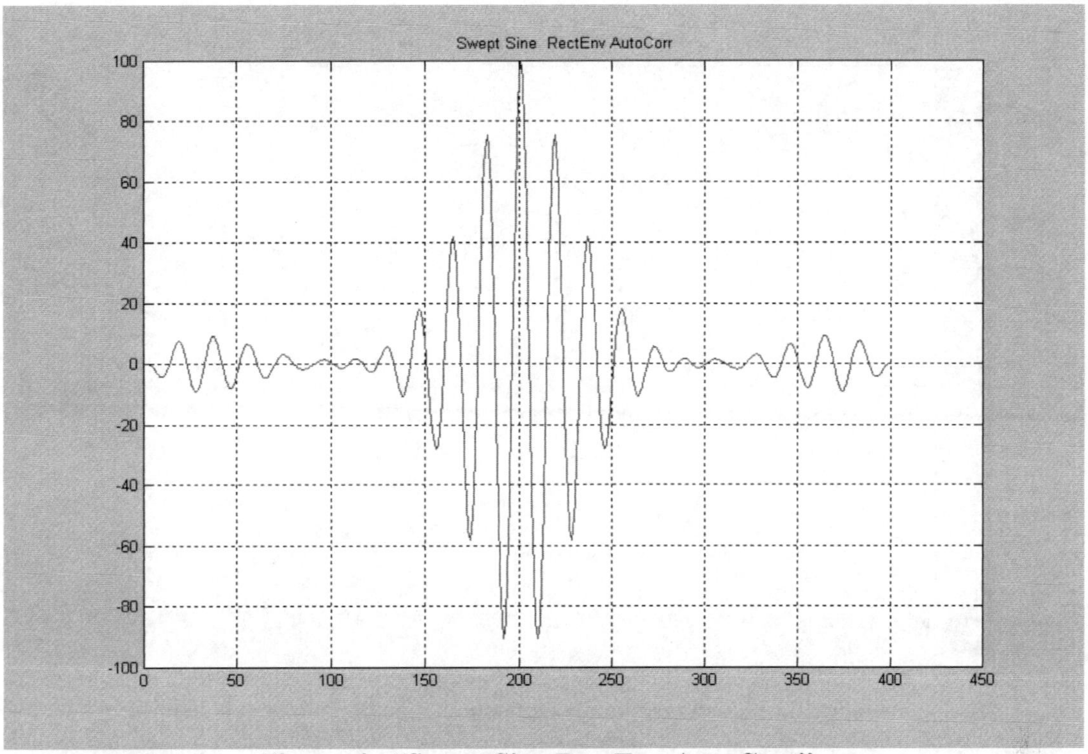

Figure 4. "Swept Sine RectEnv AutoCorr"

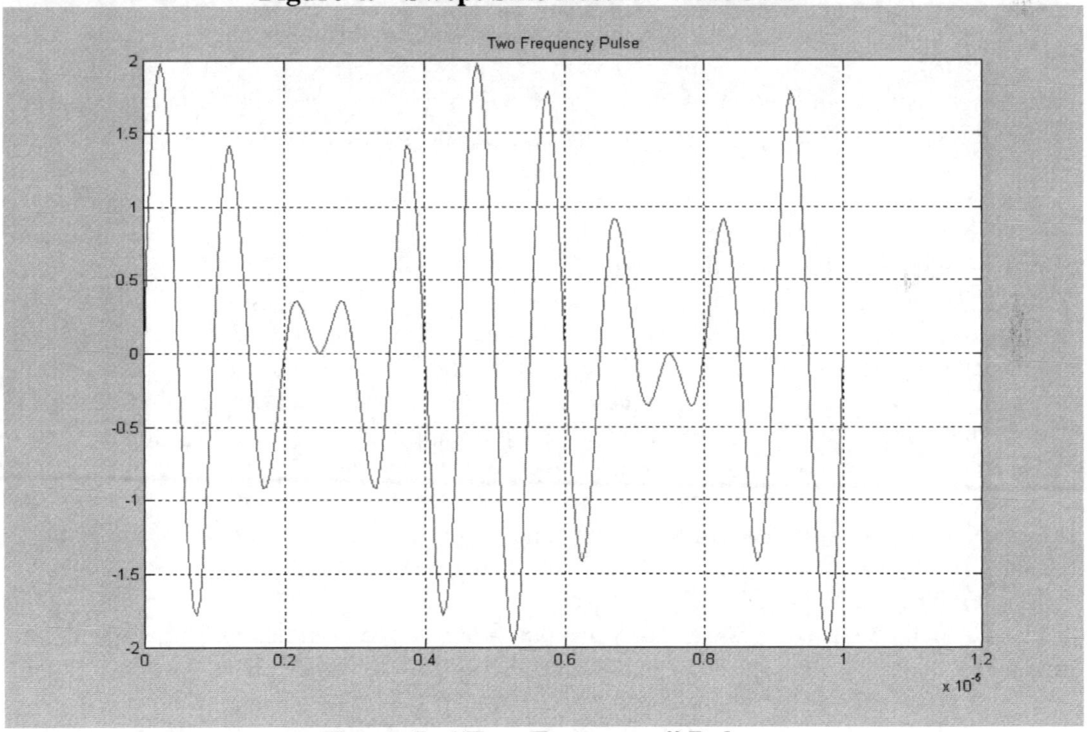

Figure 5. "Two Frequency" Pulse

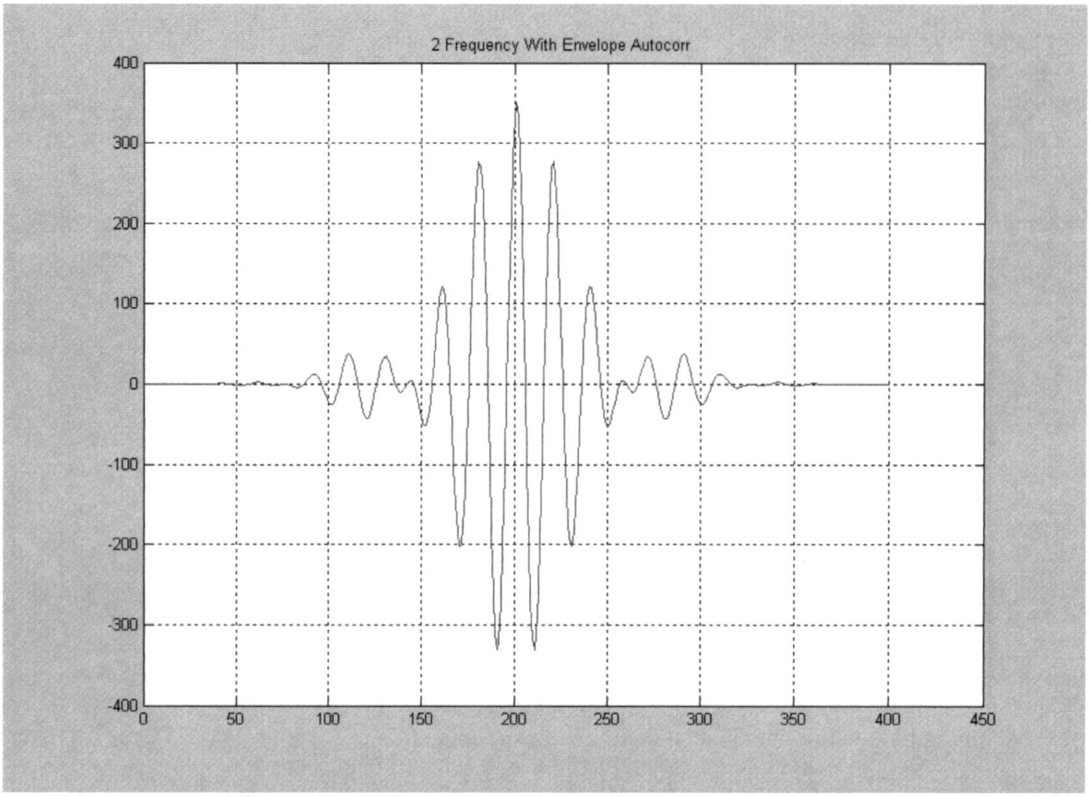

Figure 6. "2 Frequency With SinEnv Autocorr"

Another possible improvement is to generate a pulse consisting of two distinct frequencies, close to the resonant frequencies of the transducers, and added together. The two frequencies are carefully selected such that they consist of integer number of cycles each and correspond to the resonance points of the transducers. For example, if the pulse duration is 10 microseconds and the two frequencies are to have nine and eleven complete cycles each, the calculation is as follows:
$T = N1*T1 = N2*T2$ where T is the 10 microseconds, $T1 = 1/f1$ and $T2 = 1/f2$ are the periods of the two frequencies in question and N1 and N2 are the number of cycles that are to occur in the given time T. Arbitrarily choosing N1 to be 9 and N2 to be 11, it can be found that the corresponding frequencies are f1 = 0.9 MHz and f2 = 1.1111... MHz. The frequencies selected will depend on the actual application. Figure 5 illustrates this pulse and Figure 6 the autocorrelation. By combining the frequencies this way, the correlation has a "beating" effect, which improves the results by accentuating the difference in the amplitudes of the primary and secondary peaks in the cross correlation.

5. System Models

To evaluate the effectiveness of our proposed improvements and to compare with the single frequency methods used, we modeled the system and the pulses using Matlab®. The system we modeled consisted of two narrow band transducers that could each operate as both a receiver or a transmitter, and a medium, such as water, to transmit the signal. The transducers were modeled after units actually in use [2]. These transducers had a tendency to oscillate at their resonant frequency even after they were no longer energized. The model for the transducer created in Matlab® is an approximation of the real system and may not fully incorporate all of the complexities associated with an actual transducer [3]. Although the transducers are not modeled in their full complexity, they clearly illustrate the advantages of the multi frequency pulses over the single frequency ones. Our intent here is to show that by carefully designing the pulse, tangible improvement, over the single frequency method, can be realized. A block diagram of the Matlab® model used (which includes the pulse generator, the transmitter, the medium, and the receiver) is illustrated in Figure 7.

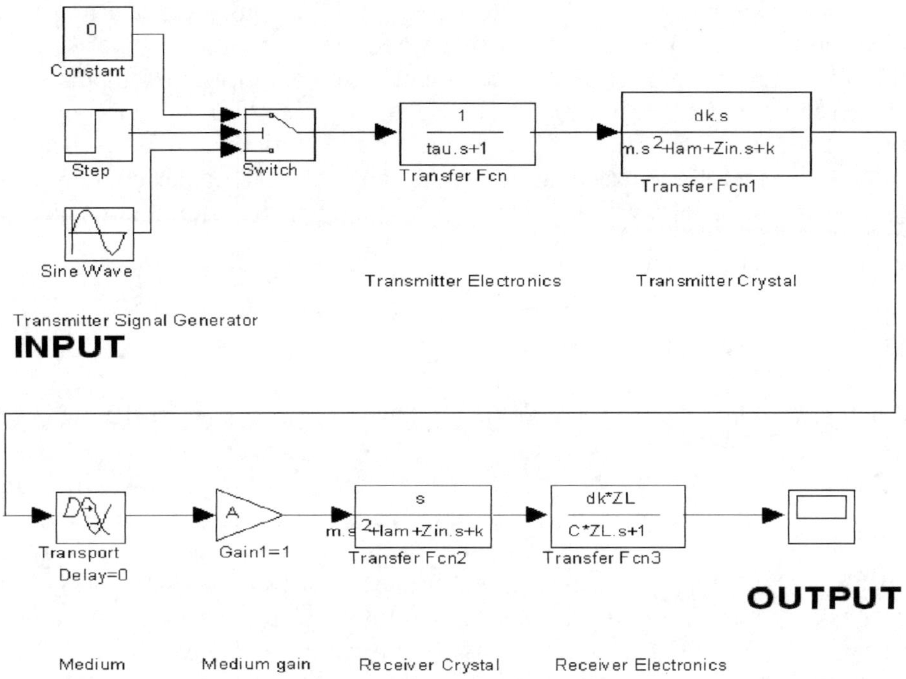

Figure 7. "System Block Diagram"

6. Testing the Pulses

In order to evaluate the efficacy of our proposed solution we generated a series of pulses with different frequency contents, envelopes and durations. We will illustrate only a small sampling of these in this paper. To obtain a baseline, we generated pulses of 10 cycles of a 1 MHz frequency of amplitude 1. This frequency was selected to match the narrow frequency bandwidth of the transducers. To illustrate a comparison of the results we did an autocorrelation of each of three types of pulses. Each was investigated both with a rectangular (no amplitude modulation) envelope and with a sinusoidal envelope. The latter envelope e(t), which we will call Sinusoidal, has the form of Equation (1).

$$e(t) = 1 - \cos(xt). \text{ For } 0 < t < 2*\pi/x. \quad (1)$$

This envelope rises and descends smoothly from a zero amplitude to a maximum value of two and returns to zero. We arbitrarily selected this envelope to match the duration of the pulse. Other envelope shapes are also possible (such as Gaussian) but the main objective here is to demonstrate the advantages of certain pulse designs based on their frequency content. We used Matlab® [4] to do the modeling and mathematical operations. To get a baseline we started with a pulse consisting of 10 cycles of a pure 1 MHz sinusoidal signal with a rectangular envelope (that is to say, the signal abruptly starts and abruptly stops) as shown in Figure 1. The time function CP(t) is depicted in Equation (2).

$$CP(t) = A*\sin((2*\pi*10^6)*t) \text{ for } 0 < t < 10*10^{(-6)} \qquad (2)$$

Doing an autocorrelation on this 10 microsecond, 1 MHz pulse reveals a "diamond" shaped waveform as shown in Figure 2. This waveform exhibits secondary peaks that have an amplitude approximately 93% of the main peak. Multiplying the pulse with the Sinusoidal envelope and repeating the autocorrelation gave similar results. That is the secondary peaks were also around 93%, however the tertiary and other peaks were considerably lower than the corresponding peaks with the rectangular envelope. In the presence of noise it becomes more difficult to identify the maximum peak of the correlation.

The second pulse design that we are considering here is also 10 microseconds in duration but consists of a linearly swept frequency ranging from 0.9 MHz to 1.1 MHz. Equation (3) and Figure 3 illustrate the waveform.

$$SP(t)= A*\sin(2*\pi*(0.9*10^6+(0.2*10^6)*t/(10*10^{(-6)}))*t),\ 0 < t < 10*10^{(-6)} \qquad (3)$$

The autocorrelation of this pulse, with the rectangular envelope shown in Figure 4, illustrated a dramatic improvement over the single frequency pulse CP(t). The secondary peaks of the autocorrelation of the swept pulse were less than 77% of the primary peak. This greatly improves the ability to discriminate the primary from secondary peaks. The swept pulse was then multiplied by the envelope and again autocorrelated. As expected the main peak was lowered and a small degradation in performance was noticed. This resulted in secondary peaks approximately 81% of the main peak. This, however, is still a great improvement over the single frequency case.

The third pulse design we considered, TF(t), given by Equation (4) and shown in Figure 5, was also 10 microseconds in duration and consisted of the sum of the two distinct frequencies, 0.9 MHz. and 1.1 MHz.

$$TF(t)= A*(\sin((2*\pi*0.9*10^6)*t)+\sin((2*\pi*1.1*10^6)*t)),\ 0<t< 10*10^{(-6)} \qquad (4)$$

With a rectangular envelope this pulse exhibited the best results with secondary peaks 64% of the primary peak. When the Sinusoidal envelope was incorporated, the secondary peaks increased to 78% as depicted in Figure 6. This proved to be the best performing of the three pulses considered.

It was then a question to see how well these pulses performed in the simulated "real" system. The three pulses, with Sinusoidal envelopes and injected noise were each applied to the input of the system and correlated with the output. The particular characteristics of the system and the noise may further degrade the performance of the pulses, but the overall performance of the multi-frequency pulses was far superior to that of the single frequency pulse. Figure 8, Figure 9 and Figure 10 illustrate these results. Figure 11 illustrates the relatively small effect that 25% noise had on the correlation.

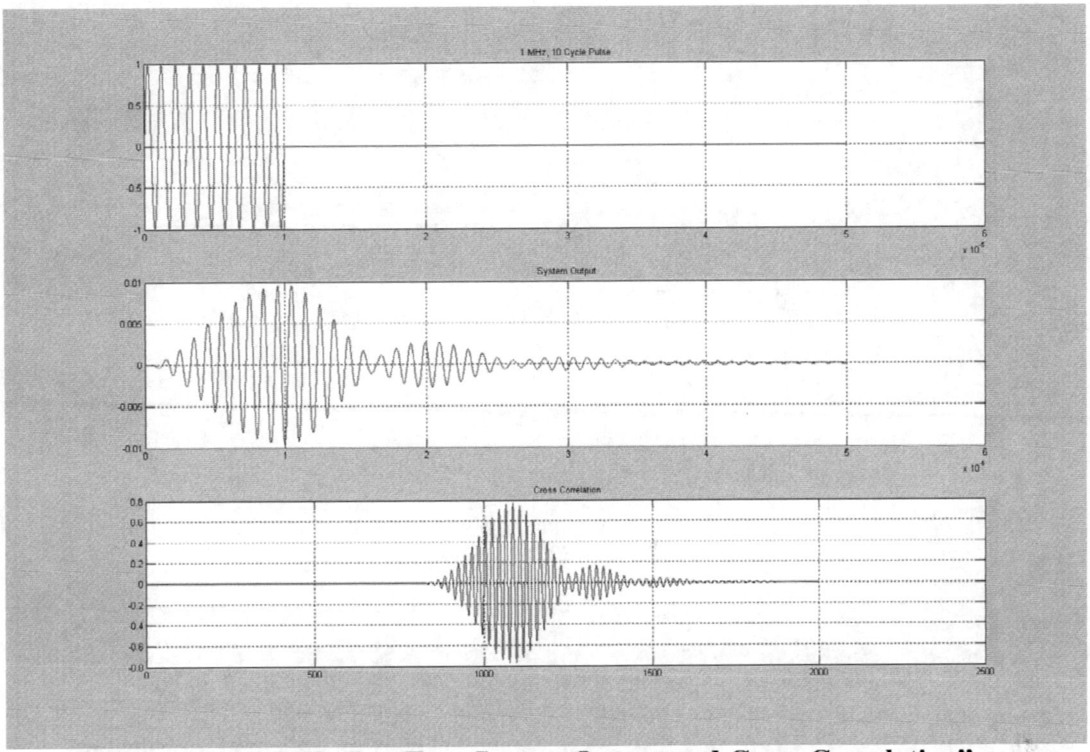

Figure 8. "1 MHz RectEnv, System Output and Cross Correlation"

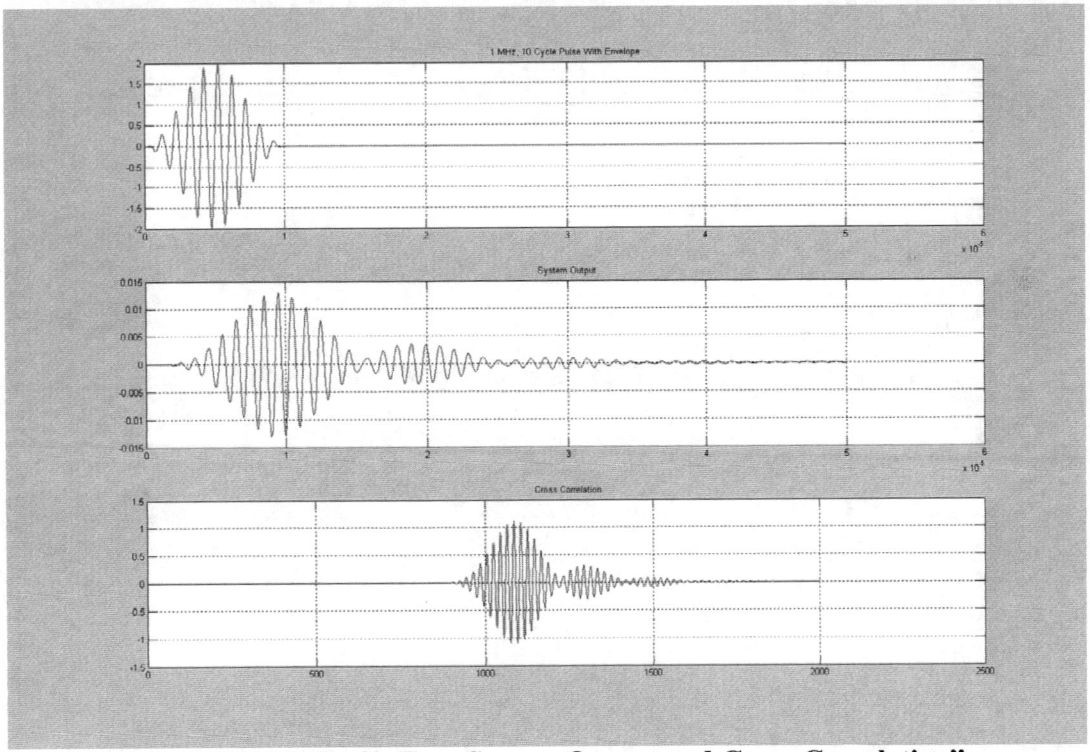

Figure 9. "1 MHz SinEnv, System Output and Cross Correlation"

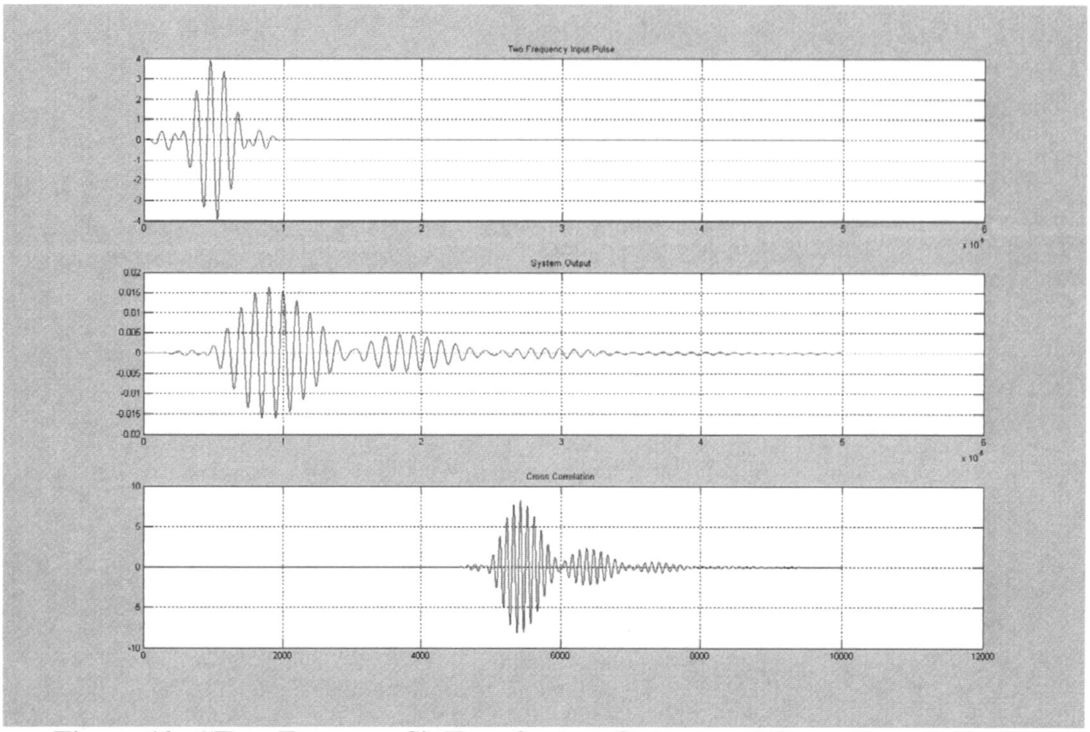

Figure 10. "Two Freqency SinEnv, System Output and Cross Correlation"

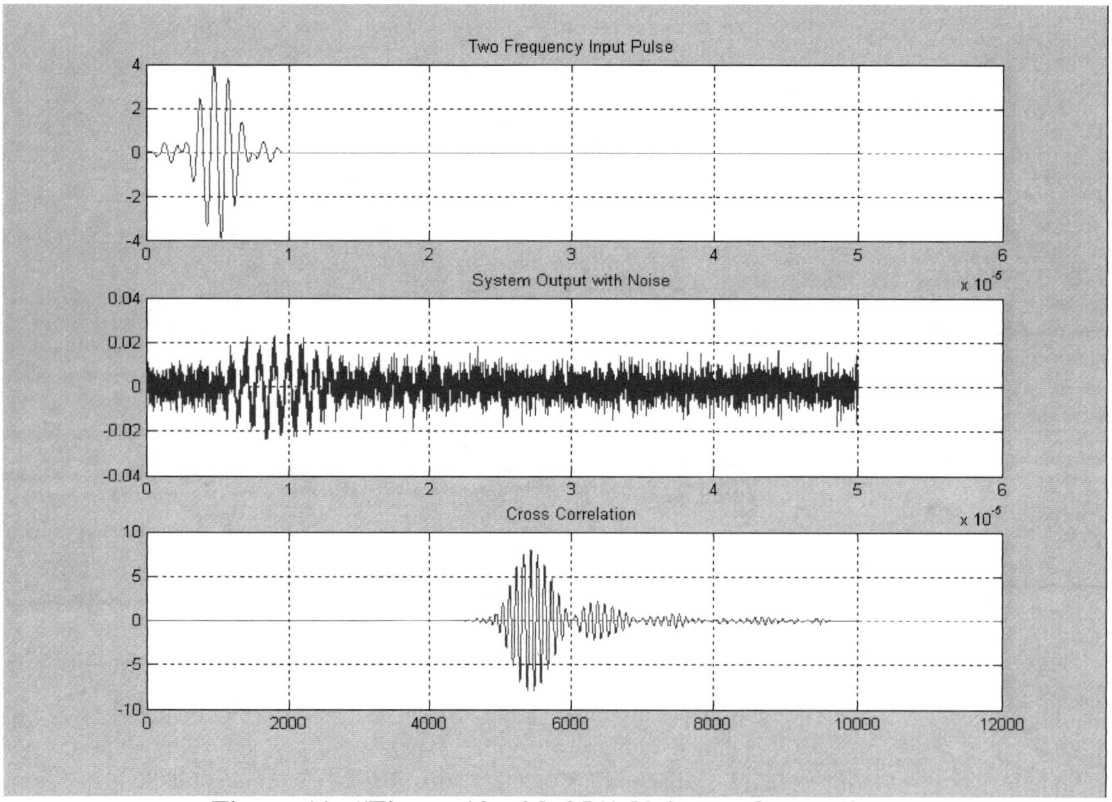

Figure 11. "Figure 10 with 25% Noise on Output"

7. Conclusions

Accurately measuring the arrival time of an ultrasonic pulse in a "real" system with nanosecond accuracy is a difficult problem. This necessitates the use of carefully crafted signals and sophisticated signal processing techniques. To maximize the accuracy of time of arrival measurement, the pulse should be customized to best operate within the system parameters. To make the transit from transmitter and be accurately recognized at the receiver, the pulse should be made as unique as possible. This is necessary because the pulse will be distorted in transit and also by the transducer characteristics. In this paper we briefly discussed the advantages of multiple frequency pulses over the currently used single frequency versions. The swept frequency pulse is fashioned after a technique used in the seismic exploration field. The application considered here, however, had the additional limitations of short time duration and high frequencies, making the pulse design even more critical. One advantage of a narrow band system is that the injection of white noise in the system has little effect on the received signal as evidenced by Figure 11. In this paper we compared the performance of the single frequency pulse with our proposed swept frequency and two-frequency pulses. Table 1. below summarizes the findings. The pulses were all 10 microsecond duration and the two envelopes used were Rectangular and Sinusoidal. An autocorrelation of each pulse was performed and a measure taken of the relative amplitudes of the primary and secondary peaks. The primary peak was considered 100%.

Table 1. "Relationship of Primary and Secondary Peaks"

10 Microsecond Pulse	Autocorr. Peak	Secondary Peak Amp.	Envelope
1 MHz.	100%	93%	Rectangular
0.9 MHz to 1.1 MHz Linear Sweep	100%	77%	Rectangular
0.9 MHz to 1.1 MHz Linear Sweep	100%	81%	Sinusoidal
0.9 MHz and 1.1 MHz Frequencies	100%	64%	Rectangular
0.9 MHz and 1.1 MHz Frequencies	100%	78%	Sinusoidal

If system parameters could be modified, i.e. if the transmitter and receiver can be made to have slightly different frequencies or if the transducer bandwidth broadened the pulse frequencies should be selected to take advantage of these facts. Methods to take advantage of different transducer designs, such as broadening the bandwidth or using different resonant frequencies in the transmitter and receiver, are being studied. It should be noted that white noise had relatively little effect on a system with such tight frequency constraints. This is so because most of the noise energy is filtered out. In summary, and based on our findings, multi frequency pulses outperformed their single frequency counterpart operating under similar conditions. Of the two pulse designs we proposed, the swept frequency pulse showed great improvement over the single frequency pulse, but the two frequency pulse was the one that performed the best.

Acknowledgements

The authors gratefully acknowledge the contributions provided by Roderick Melnik for his encouragement and Morten Willatzen for providing the problem and his help in developing some of the solutions discussed in this paper. Jonathan McAfee's editorial suggestions were greatly appreciated. We are also grateful to all involved in making EMCI47 a most educational and productive experience.

References

[1] Baeten, G. and Ziolkowski, A., *The Vibroseis Source*, Elsevier Science Ltd., 1990.

[2] M. Willatzen, "Ultrasound Transducer Modelling, General Theory and Applications to Ultrasound Reciprocal Systems", IEEE Trans. Ultras., Ferroelectrics, and Freq. Control, 48, 100-112, Jan. 2001.

[3] M. Willatzen et al., "Arrival-Time Detection and Ultrasonic Flow-Meter Applications", (This Journal).

[4] Matlab, The Mathworks, Inc., MA, USA.

[5] Defatta, D., Lucas, J., and Hodgkiss, W., *Digital Signal Processing: A System Design Approach*, John Wiley and Sons, Inc., 1988.

[6] Phillips, C., and Parr, J., Signals, Systems, and Transforms Second Edition, Prentice-Hall, Inc., 1995.

[7] Ziemer, R., Tranter, W., and Fannin, D., Signals & Systems Continuous and Discrete, Prentice-Hall, Inc., 1998

Institute of Physics Publishing
doi:10.1088/1742-6596/52/1/003
Journal of Physics: Conference Series **52** (2006) 27–35
Mathematics for Industry in Denmark

Distance of Sample Measurement Points to Prototype Catalog Curve

Poul G. Hjorth[1] **and Mirza Karamehmedović**[2]
Department of Mathematics, Building 303, Technical University of Denmark, DK-2800 Kgs. Lyngby, Denmark.

John Perram[3]
Maersk-Mcinney Institute for Production Technology, University of Southeren Denmark, DK-5230 Odense, Denmark.

Robert Piché[4]
Department of Mathematics, Tampere University of Technology, FIN-33101 Tampere, Finland.

Ulrik Ullum[5]
Research and Technology, Grundfos Management A/S, DK-8850 Bjerringbro, Denmark.

Michael Krätzschmar[6]
Flensburg University of Applied Sciences, D-24943 Flensburg, Germany.

Abstract.
We discuss strategies for comparing discrete data points to a catalog (reference) curve by means of the Euclidean distance from each point to the curve in a pump's head H vs. flow Q diagram. In particular we find that a method currently in use is inaccurate. We propose several alternatives that are accurate, fast, robust and stable.

Keywords: Geometry, orthogonal distance from point to curve, Schwarz-Christoffel transformation, cubic splines.
PACS: 02.30 Approximation Methods
MSC: 51.1 Geometry. 65.1 Numerical algorithms.

1. The Working Group

This problem was worked on by Jiri Felcman, Poul Hjorth, Mirza Karamehmedović, Michael Krätzschmar, John Perram, Robert Piché, and Ulrik Ullum.

[1] Corresponding author: p.g.hjorth@mat.dtu.dk
[2] M.Karamehmedovic@mat.dtu.dk
[3] jperram@mip.sdu.dk
[4] robert.piche@tut.fi
[5] ullum@ullum.dk
[6] kraetzschmar@mathematik.fh-flensburg.de

2. The Problem

The company GRUNDFOS compares production samples of a pump to a prototype of the pump. For the prototype, a so called *reference curve* has been constructed in a (Q, H)-diagram, by taking a large number of measurements and subsequently fitting a polynomial of degree at most 5 to these data points. The shape of the reference curve in the (Q, H)-diagram reflects the classical Bernoulli relation $H \propto Q_0^2 - Q^2$ where the pressure head is related to the square of the flux Q. The design point is about half way between Q_{min} and Q_{max}.

The comparison with production samples consists of placing measurement points from samples in the (Q, H)-diagram along with the reference curve, and then for each sample and each measurement point to find the *orthogonal distance from the measurement point to the reference curve.* See figure 1.

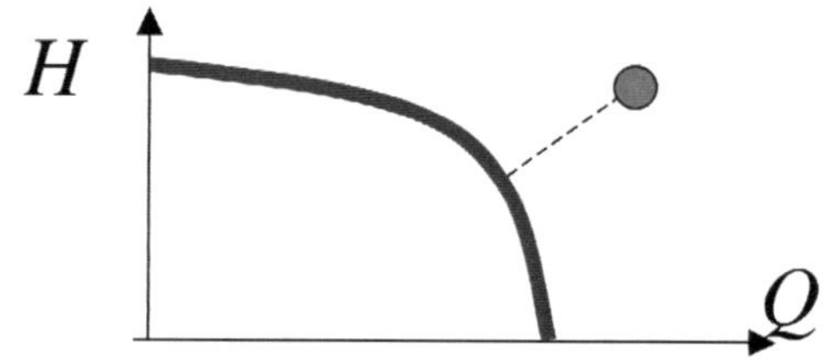

Figure 1. In the (Q, H)-plane the distance between each single measurement point and the reference curve is to be evaluated.

The details of the method currently used is given in the next section. GRUNDFOS asks for a suggestion for an optimal distance-checking procedure which is sufficiently self contained, e.g., not relying on third-party packages or numerical libraries.

3. The Present Method

The method currently being employed at GRUNDFOS is in more details as follows. To begin with, the catalog curve is approximated by a piecewise linear curve. An algorithm (termed 'Inverse Schwarz-Christoffel Transformation', see however the next section) is being used to find the distance of the measurement points to the reference curve (Figure 2).

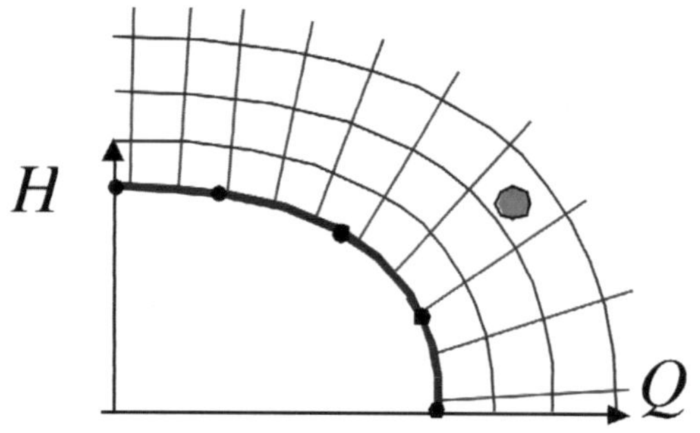

Figure 2. The procedure currently used essentially places a grid near the point and the curve and evaluates the position of the point in the coordinates corresponding to the grid.

For each measurement point, a new coordinate system (x', y') is introduced such that the y'-axis is the line through the point and perpendicular to the piecewise linear segment near the measurement point. The x'-axis is then the the line through this segment, and the distance from the point to the curve is read off as the y'-coordinate of the point. The idea stems from the theory of conformal mappings where it forms a part of the so called Schwarz-Christoffel Transformation. GRUNDFOS accordingly terms their algorithm the "Inverse Schwarz-Christoffel Transformation" (ISC).

There is, however a nonuniqueness problem associated with this method (apart from the fact that it is somewhat computationally cumbersome.) Before suggesting alternative methods, we examine this problem in more detail.

4. The Schwarz-Christoffel Transformation

It is useful firstly to provide a brief recapitulation of the Schwarz-Christoffel transformation.[7] Given a finite index set $I = \{1, \ldots, n\}, n \in \mathbb{N}$, let $\{x_j\}_{j \in I} \subset \mathbb{R}$ be a set of points on the real axis. Furthermore, introduce the set $\{\theta_j\}_{j \in I} \subset \mathbb{R}$, as well as the constants $A, B \in \mathbb{C}$. A Schwarz-Christoffel transformation is now defined by (see [1])

$$w(z) = A \int_0^z \prod_{j=1}^{n} (\zeta - x_j)^{\theta_j} \, d\zeta + B, \quad z \in \mathbb{C}, \Im z \geq 0 \tag{1}$$

It can be shown [1] that the mapping in 1 takes the segments

$$]-\infty, x_1[\,,\ ]x_1, x_2[\,,\ \ldots,\ ]x_{n-1}, x_n[\,,\ ]x_n, \infty[\tag{2}$$

of the real axis into linear segments of a polygonal path in the complex plane such that the 'right turn angles' between the successive segments are specified by $\theta_j, j \in I$. In the special case where it holds true that

$$\lim_{z \to -\infty} w(z) = \lim_{z \to \infty} w(z) \tag{3}$$

the upper half-plane is mapped onto a polygon comprising the $n+1$ vertices

$$w(x_1), \ldots, w(x_n), \lim_{z \to \infty} w(z) \tag{4}$$

In fact, given any positively oriented polygon with consecutive corners $w_j \in \mathbb{C}$ and right-turn angles $\theta_j \in \mathbb{R}$, there exists [1] an injective Schwarz-Christoffel mapping w from the upper half-plane onto that polygon such that $w(x_j) = w_j, j \in I$ and $\lim_{z \to \infty} w(z) = w_{n+1}$.

Two major comments are in order regarding the use of the ISC mapping for measurement of distances of points from curves. Firstly, a Schwarz-Christoffel transformation generally maps a rectangular coordinate system in the complex plane to a skew coordinate system in that plane. This renders the mapping inadequate for measurements of orthogonal distances of points from curves, as will be illustrated below using a practical example. Secondly, it is clear from the expression 1 that the transformation (and hence also its inverse) is dependent upon the specific choice of the right-turn angles $\theta_j, j \in I$. Expressed differently, every distinct shape of the given piecewise linear curve (catalog curve) will result, in general, in a distinct ISC mapping. This is expected to render the method sensitive to inaccuracies in catalog curves, and a numerical example given below serves to confirm this point.

A study of the algorithm employed by GRUNDFOS reveals that the company, in fact, is not using the inverse Schwarz-Christoffel transform, as asserted in the problem formulation. It is, however, still of interest to explore the possibility of an application of this method in the present context.

[7] We direct to Reference [1] for a more thorough treatment.

5. Numerical Examples

In this section, numerical examples are provided which illustrate the inadequacy of the ISC transform in the context of determination of the shortest distance of a point to a catalog curve. For completeness, we note that the particular ISC map employed here is specified by the (convenient) choice of parameters reading $(x_1, x_2) = (-1, 0)$.

In the following, a part of a catalog curve is modeled using a piecewise linear approximation, shown in Figure 3a). In the same figure, two points are inscribed such that they have the same shortest distance

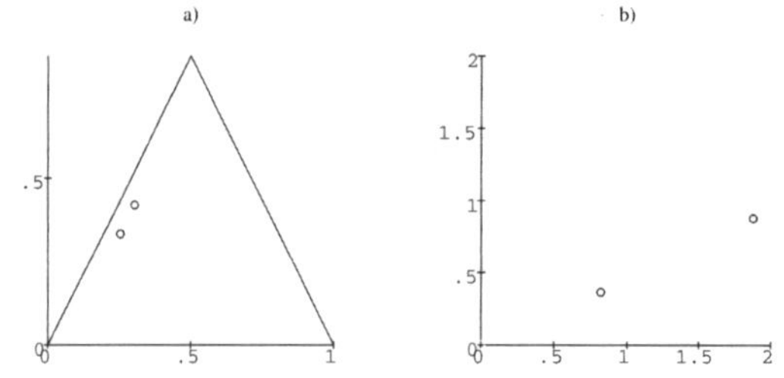

Figure 3. Inverse Schwarz-Christoffel mapping of two distinct points equidistant with respect to the data curve.

to the piecewise linear curve. Figure 3b) shows the images of these two points under the ISC mapping. As is evident, these images are points in the complex plane with different shortest distance from the real axis. Thus, the ISC transform is not suitable for direct measurement of shortest distances of points from curves.

We next consider two distinct piecewise linear curves, as shown in Figure 4. The smallest angles

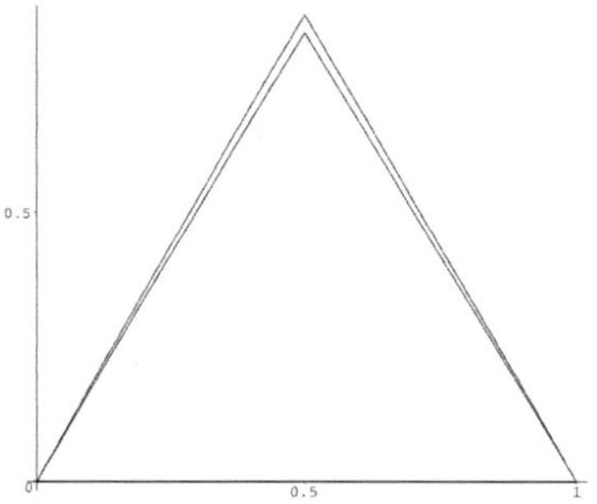

Figure 4. A model example of a simple discrepancy between two data sets.

between the linear segments of the curves are 60° and 62°, respectively. The discrepancy between the curves introduced here may model the effect of inaccuracies in catalog curve representation (piecewise linearisation.) Alternatively, the two piecewise linear curves may simply be assumed to represent two different catalog curves.

The two graphs are shown again in Figures 5a) and 6a), respectively, along with sets of points chosen such that the shortest distances from corresponding points to the respective curves are the same.

Figures 5b) and 6b) show the images of these points under the ISC transform.[8] An investigation of the presented results reveals that the shortest distances of corresponding image points to the real axis are not the same. Furthermore, we find that the ratio of the distances from the real axis of image points in case a) is not the same as the corresponding ratio of case b), the ratios being 0.484 and 0.486, respectively.

[8] Note that, for convenience, the ordinates in figures 5b) and 6b) start at 0.8 instead of at 0.

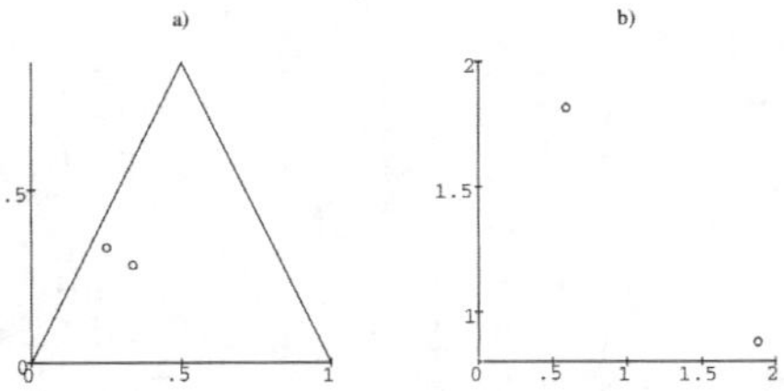

Figure 5. The ISC mapping of two points for the first data set.

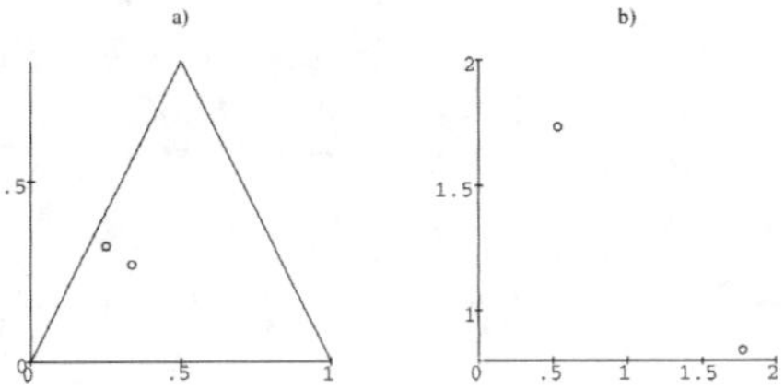

Figure 6. The ISC mapping of two points for the second data set.

Hence, a slight change in the catalog curve data evidently results in a different evaluation of the shortest distances and their ratios. This is clearly a disadvantage of the method under investigation.

As a final example, consider Figure 7a), where the original piecewise linear curve (comprising the angle of 60°) is presented together with two pairs of points. One pair of points is obtained using a linear

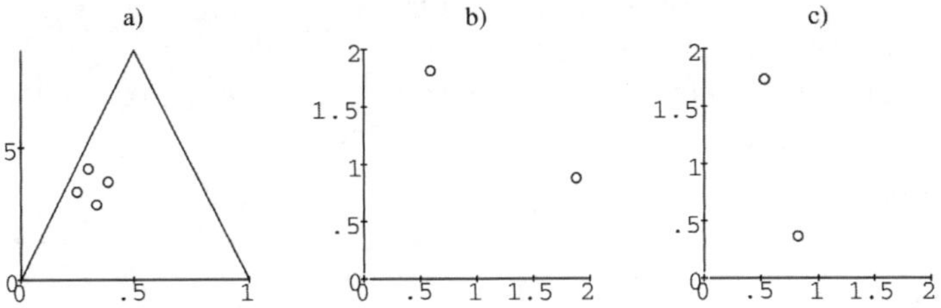

Figure 7. Illustration of change of shortest distance ratio due to simple translation of measurement and reference data.

translation of the other one in such manner that both shortest distances to the curve are preserved. Figures 7b) and 7c) show the images of the two point pairs under the ISC transform. It is seen that the shortest distances to the curve, as well as the ratio of these distances, are altered by the simple translation procedure described above.

6. Proposed Alternative Algorithms

In light of the examples of the previous section we conclude that the ISC method used currently has some serious drawbacks; it is both inaccurate and inconsistent. We therefore proceed to suggest four alternatives: Discretization, the Golden Ratio Search, and Rootfinding (Brent's Method and Sturm Sequence).

The proposed alternative methods are all minimisation algorithms. For a given continuous catalog curve $Q \mapsto (Q, H(Q))$ and a test point $P = (Q_P, H_P)$, the problem amounts to locating the minimum of the squared distance function

$$F(Q) = (Q - Q_P)^2 + (H(Q) - H_P)^2 \tag{5}$$

Since F is continuous it attains its minimum somewhere in its domain $[0, Q_{max}]$. More than one point on the curve may however realise the minimum, though this does not generically occur in practice. However, the minimum distance value is unique, so the proposed methods do not suffer from the inconsistencies of the ISC method.

6.1. Discretization

This method uses a 'brute force' approach. From a table of N catalog curve points $Q_1, \ldots, Q_N$, a table of N distance values $F(Q_1), \ldots, F(Q_N)$ is generated; the computer then picks the smallest of these value. This method is simple, but is sufficiently fast, accurate and reliable for this problem.

Suppose the search locates $F(Q_k)$ as the minimum value, and let $F(Q^*)$ be the real minimum value of the function F (assumed twice differentiable). From Taylor's theorem we then have

$$|F(Q_k) - F(Q^*)| \leq \max|F''(Q)|\max|Q_i - Q^*|^2 \tag{6}$$

so that error decreases quadratically with the density of curve points.

6.2. Golden Ratio Search

In this standard minimisation algorithm, a succession of minimum-bracketing Q intervals, starting with the domain endpoints $[a, b] = [0, Q_{max}]$, is generated as follows. From the two values a, b compute the intermediate values $c = a + 0.382(b - a)$ and $d = a + 0.618(b - a)$. If $F(c) \leq F(d)$ then continue the search in $[a, d]$, otherwise continue the search in $[c, b]$. The locations of the intermediate points are such that one of the intermediate values can be reused in the next iteration, so once the method gets started only one evaluation of F per iteration is needed. This algorithm gives quadratic convergence to a local minimum of F when F is unimodal – this is typically the case for distance functions considered here.

6.3. Root-Finding

The minimum of F is achieved either at the end points of the domain $[0, Q_{max}]$ or at an interior point. In the latter case, we should locate the zeros of the derivative F'. Because the catalog curve function $H(Q)$ is a polynomial of degree $n \leq 5$, the squared distance function F given by Eqn (5) is a polynomial of degree $2n = 10$, and the derivative is a polynomial of degree $2n - 1 \leq 9$.

There are many methods available to find a single function zero. A popular choice is Brent's method, also known as the van Wijngaarden-Deker-Brent method [7]. This algorithm combines root bracketing, interval bisection, and inverse quadratic interpolation, and does not require function derivatives. The algorithm is described in a number of textbooks, including [5] and [7].

A general root-finding algorithm may find a solution that corresponds to a local maximum or to a local minimum that is not a global minimum. To ensure against this, one could factor the computed zero out of the polynomial F' ("deflation") and restart.

Alternatively, there are methods available to reliably find all the real zeros of a real polynomial. For example, reference [2] gives an easily programmable algorithm that uses Sturm's sequences [6] to bracket the roots, and subsequently uses bisection to pinpoint the zeros.

7. Reference Curves and Tolerances

The problem brought to us by the company is a straightforward geometry question, but the underlying problem is more involved: how to evaluate how well a sample meets a tolerance, and how this tolerance is determined. We had too little background information to be able to do any detailed work on this question, but we have some general suggestions that could be the basis of further study.

7.1. Reference Curve Fitting

The reference curve is constructed based on measurements of a number of standard pumps. Orthogonal distance is used as a curve-fitting criterion. The public domain code ODRPACK[4], is available to compute polynomial reference curves based on orthogonal distance regression. This code could be useful in automating the reference curve construction.

Orthogonal distance regression is also known as error-in-variables modelling in statistical literature. Standard statistical regression tools are available to help determine the appropriate degree of the polynomial and to provide confidence regions above and below the reference curve. These confidence regions might provide a better way of evaluating sample pumps than the geometrically motivated tolerance curves currently used (see below).

7.2. Splines

Instead of fitting a single polynomial, one could construct a reference curve that is a piecewise cubic polynomial with continuous first and second derivatives, that is, a cubic spline. This would generally produce a better fit to measurement data while at the same time limiting the degree of the polynomial. (High degree polynomials are more likely to have undesirable oscillations and make it harder to compute the zeros of the derivative function F' in the Root-Finding method.)

When using cubic splines the interval $[0, Q_{max}]$ is divided into m (not necessarily equally spaced) subintervals $[Q_i, Q_{i+1}[$. The cubic spline restricted to any subinterval is a polynomial of degree 3. All together, these m polynomials of degree 3 form a function that is a piecewise polynomial and furthermore twice continuously differentiable. Because the eye cannot distinguish between a twice continuously differentiable and an infinitely continuously differentiable function there is no aestethic reason not to prefer a cubic spline instead of a polynomial as a catalog curve. ODRPACK could be used for automatic fitting of cubic splines to measurement data based on orthogonal distance regression.

Using cubic splines instead of a single polynomial curves would change the Root-Finding Method for finding the distance between a given point and the catalog curve in the following way. Instead of looking for all the zeros of a polynomial of degree $2n-1$ in the interval $[0, Q_{max}]$, we find for each subinterval $[Q_i, Q_{i+1}]$ all the zeros of a 5th order polynomial belonging to this interval. Let $\mathcal{F}$ denote the set of all the values of F at these zeros together with the values of F at node points $Q_1, \cdots, Q_m$. Then $\mathcal{F}$ is finite with no more than $6m+1$ elements and the continuous minimization problem can be again be replaced by the discrete minimization problem. If the cubic spline is a convex or concave function (which will typically be the case for a catalog curve) the number of points in $\mathcal{F}$ will be close to m instead of being close to $6m+1$.

7.3. Tolerance curves

GRUNDFOS defines tolerance curves by adding $\pm 7\%$ to the H values and $\pm 9\%$ to the Q values, thereby generating 4 curves. There are two displaced set of H values plotted against Q, and two sets of Q values displaced from the catalog curve. See figure 8.

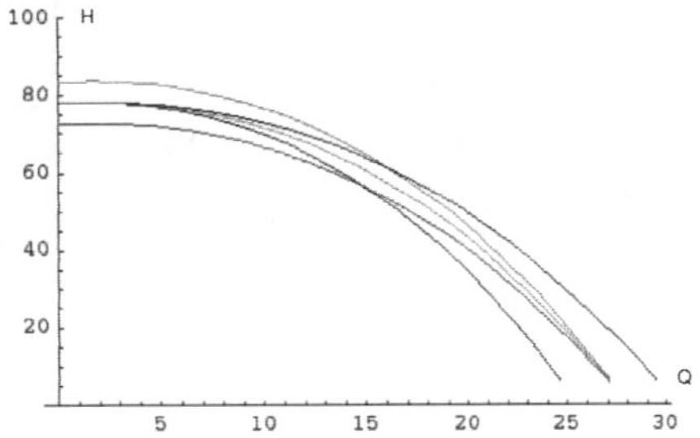

Figure 8. The 'tolerance curves' obtained by horizontal and vertical displacement of the catalog curve.

Since the tolerance of samples is measured in terms of perpendicular distance of a test point to the catalog curve, it is far from clear that the curves drawn in the above manner are consistent with that measure. To resolve this, we construct curves that are obtained from the catalog curve by displacing the curve in directions normal to the tangent at each point.

The amount displaced we chose to be equal in both directions and of magnitude the r.m.s. error

$$\frac{1}{100}\sqrt{(9Q)^2 + (7H(Q))^2} \tag{7}$$

The normal direction vector is given by

$$\mathbf{n} = \frac{(-H'(Q), 1)}{\sqrt{1 + H'(Q)^2}} \tag{8}$$

Thus, for the point $(Q, H(Q))$ on the curve, the corresponding curve on the upper and lower tolerance curve have the parametric experssions:

$$(Q, H(Q)) \pm \frac{1}{100}\sqrt{(9Q)^2 + (7H(Q))^2}\frac{(-H'(Q), 1)}{\sqrt{1 + H'(Q)^2}} \tag{9}$$

The resulting curve (for a typical shape of the catalog curve) is shown superimposed on the GRUNDFOS curves in figure 9.

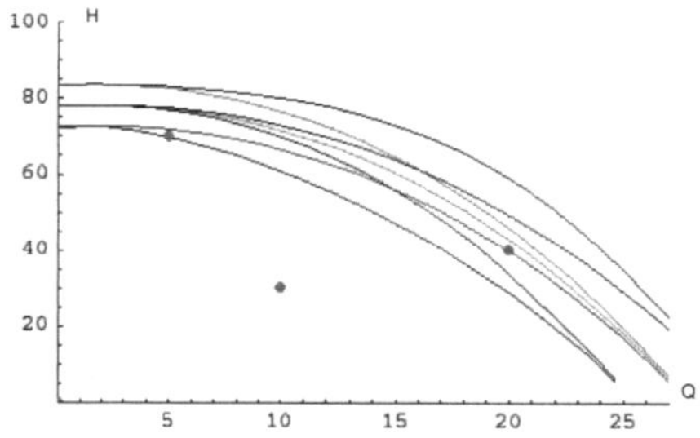

Figure 9. The tolerance curves obtained by displacement of the catalog curve along the direction in which tolerance is measured. Note that the curve is different from the envelope of the curves of figure 8.

Notice that the upper curve generated in this way (which is consistent with the calculation of the sample distance values) lies above the horizontal-vertical shifted curves. Similarly, the lower curve lies below the horizontal-vertical shifted curves. The differences are quite significant near the duty point, indicating that the tolerance curves used up to now have been too stringent.

8. Conclusion

In this report we have first given reasons GRUNDFOS should change the procedure for comparison of sample measurements with the catalog curve; indeed the procedure currently used may give misrepresenting and misleading results. We have then proposed four separate algorithms to solve the problem.

Of these four algorithms, we recommend *Discretization* for its simplicity, speed and robustness. The error is easily controllable through the number of points in the discretization.

[1] E. B. Saff and A. D. Snider, 1993 *Fundamentals of Complex Analysis for Mathematics, Science and Engineering*, 2nd edition, Prentice Hall

[2] See, e.g., `http://www.acm.org/pubs/tog/GraphicsGems/gems/Sturm/main.c`

[3] I.J. Anderson, M.G. Cox, A.B. Forbes, J.C. Mason and D.A. Turner 1998 *An Efficient and Robust Algorithm for Solving the Foot Point Problem*, in Mathematical Methods for Curves and Surfaces II, Dæhlen et al (eds.) Vanderbilt Press, ISBN 0-8265-1315-8 .

[4] Paul T. Boggs and Janet R. Donaldson, *ALGORITHM 676, ODRPACK: Software for Weighted Orthogonal Distance Regression*, ACM Transactions on Mathematical Software.

[5] Brent, R.P. 1973 Ch.3-4 in *Algorithms for Minimization Without Derivatives.* Englewood Cliffs, NJ: Prentice-Hall

[6] Forsythe, G.E.; Malcolm, M.A.; and Moler, C.B. 1977 7.2 in Computer Methods for Mathematical Computations. Englewood Cliffs, NJ: Prentice-Hall

[7] Press, W.H.; Flannery, B.P.; Teukolsky, S.A.; and Vetterling, W.T. 1992 *Numerical Recipes in FORTRAN: The Art of Scientific Computing*, 2nd ed. Cambridge, England: Cambridge University Press

Prediction of rotating stall within an impeller of a centrifugal pump based on spectral analysis of pressure and velocity data

Ulrik Ullum[1], Jennifer Wright[2], Oguz Dayi & Ali Ecder[3], Azzeddine Soulaimani[4], Robert Piché[5] and Hemant Kamath[6]

[1] R & T, Grundfos Management A/S, Bjerringbro, Denmark, [2] University of Bath, United Kingdom, [3] Bogazici University, Turkey, [4] Ecole de Technologie Superieure, Canada, [5] Tampere University of Technology, Finland, [6] University of Southern Denmark, Sønderborg, Denmark.

E-mail: [5] `robert.piche@tut.fi`

Abstract. Experimental data, which was acquired in two centrifugal pumps and provided by Grundfos A/S, were analysed to determine if rotating stall could be detected from the velocity and pressure time series. The pressure data, which were uniformly acquired in time at high sample rates(10 kHz), were measured simultaneously in four adjacent diffuser channels just downstream of the impeller outlet. The velocity data, which were non-uniformly sampled in time at fairly low rates(100 Hz to 3.5 kHz), were acquired either in or downstream of the impeller. Two different methodologies were employed for detection of stall. The first method, which involved direct analysis of raw data, yielded qualitatively useful flow reversal information from the time series for the radial velocity. The second approach, which was based on power spectrum analysis of velocity and pressure data, could detect the onset and identify the frequency of rotating stall to a satisfactory extent in one of the two pumps. Nearly identical stall frequencies were observed in both velocity and pressure power spectra and this rotating stall phenomenon, which occurred at a very low frequency relative to the impeller speed, did not reveal any noticeable degree of sensitivity to the flow rate. In the other pump, where the available data was limited to velocity time series, the power spectrum analysis was successful in detecting stationary stall for a 6 bladed impeller but did not provide conclusive results for the existence of stall in the case of the 7 bladed impeller. Recommendations on the type of experimental data required for accurate detection of stall are provided based upon the present study.

Keywords: Time series, rotating stall, centrifugal pump.
PACS: 06.30Ft, 47.80.
MSC: 76-05, 37M10.

1. Introduction

When a pump operates at reduced flow rates, its performance could get degraded by the inception of flow instabilities such as stationary and rotating stall phenomena. Stall refers to zones of recirculating fluid within some flow passages of a pump's components(e.g. impeller, diffuser or volute) and this can cause substantial velocity and pressure fluctuations which can not only adversely affect the flow but the pump as well. These zones could propagate along the circumferential direction of the impeller or diffuser and it is then referred to as rotating stall. The rotating-stall mechanism is not yet very well understood, even though a qualitative explanation of propagation, based upon cascade theory, was proposed by Emmons et al. [1] more than 40 years ago.

Although rotating stall may occur in any turbomachinery component, this phenomenon is most frequently studied in compressor rotor-blade passages. In centrifugal pumps, experimental studies([2]-[4]) indicate the existence of rotating stall in the impeller and in the diffuser/volute and in both for some geometries. Computational studies of stall prediction in industrial centrifugal pumps are limited primarily due to the difficulties in modelling turbulent separated flows in complex geometries that involve interactions between rotating and stationary components. A combined experimental and computational study carried out recently([5]-[6]) reveals that upstream influence of stationary stall can introduce a swirl component in the inlet velocity profile. This upstream influence can adversely affect predictions from large eddy simulations(LES), which did not account for the inlet distortion since it was not possible to experimentally measure it. It was also shown that in contrast to LES, Reynolds-averaged Navier-Stokes(RANS) simulations did not even predict the existence of stationary stall phenomenon with either of the two turbulence models which were tested in this study. Accurate computational prediction of stall is largely dependent on user-specified input such as boundary conditions, transition and turbulence modeling, computational domain that needs to be defined for a realistic representation of the actual configuration among others, which depend upon details that often can be provided only by conducting experiments.

Reliable detection of stall in industrial centrifugal pumps can then be more directly based on actual experimental measurements. Furthermore, different experimental techniques of varying complexity can be employed in order to furnish data suited for stall detection. Evidently, the simplest and the least time consuming, but yet reliable, technique is preferred.

2. Formulation of the Problem

The primary objective of this study was to develop a simple method that predicts the presence of rotating stall within an impeller of a centrifugal pump based upon time series or reduced data sets of measured mean-velocity and pressure data. Experimental data acquired in two pumps, CR90 and CR4x2, were available for analysis. For the CR4x2 pump, data was acquired for two different configurations, namely, with a 6-bladed and a 7-bladed impeller. The methodologies to be proposed for stall detection will be closely related to the nature of the available experimental data. Therefore a brief introduction to the measurement techniques and equipment used in this study, shall be given hereafter.

Laser Doppler Velocimetry (LDV) is a non-intrusive optical technique for measuring velocities in transparent fluids. An intereference pattern is obtained by intersecting two coherent laser beams and neutrally-buoyant particles, which are seeded into the flow, will scatter a signal when passing the beam intersection. The emitted signal is processed by the LDV system and converted to a signed velocity. Due to the stochastic arrival times of the particles passing through the measurement volume, the velocity time series obtained will have non-uniform time intervals between samples. The measurement system used for both the CR4x2 and the CR90 pumps is a two velocity-component back-scatter system from Dantec Dynamics with a 300mW air cooled Argon-ion laser and two BSA enhanced processors.

Apart from a transparent fluid, the LDV technique also requires a fully transparent flow rig in the regions of interest. Fig. 1 shows two images from the CR4x2 setup. It can be seen that the front and back plate of the pump housing, the hub and shroud of the impeller and the diffuser front plate are all made of either perspex or glass. This construction enables LDV measurements in a large part of the pump housing. The inlet and outlet pipes are also equipped with glass windows allowing for measurements of the velocity profiles at these locations.

On the CR90 pump, the optical access has been limited due to a very complex geometry and LDV measurements are only possible through windows or prisms in the geometry as it can be seen in the CAD illustration (Fig. 2).

In the CR4x2 pump, the velocity data used in the present study were acquired within the rotating impeller at a radial position of $0.9R_{imp}$, where R_{imp} denotes the impeller radius, and pressure data are not available. In this case, the sampling is phase-resolved i.e. velocity samples were sorted into

Figure 1. Images of CR4x2 test pump and LDV measurement optics

Figure 2. CAD illustration of CR90 pump and its optical windows for data acquisition

360 angular bins based on their arrival time relative to a once-per-revolution pulse. In each bin, mean velocity vectors were calculated based on approximately 500 individual samples. It should be noted that this bin-averaging technique can be employed for diagnosing stationary but not rotating stall.

However, in the CR90 pump, LDV data is acquired immediately downstream of the impeller (see Fig. 2). Pressure data, which were uniformly sampled in time, were acquired simultaneously in 4 adjacent diffuser-channel entrances just downstream of the impeller outlet. For pressure measurements, four fast-response Kulite pressure transducers have been used in conjunction with an MGC amplifier and a four-channel Tektronix digital oscilloscope for data acquisition. This completes a brief description of the measurement techniques employed in the present study.

The performance curves of the CR90 pump and CR4x2 pumps, with 6 and 7 bladed impellers, are shown in Fig. 3. It can be observed that the CR90 performance curve reveals a distinct loss of smoothness below a flow rate of 40 m^3/h and as a result would indicate a likelihood of onset of stall. In contrast, the

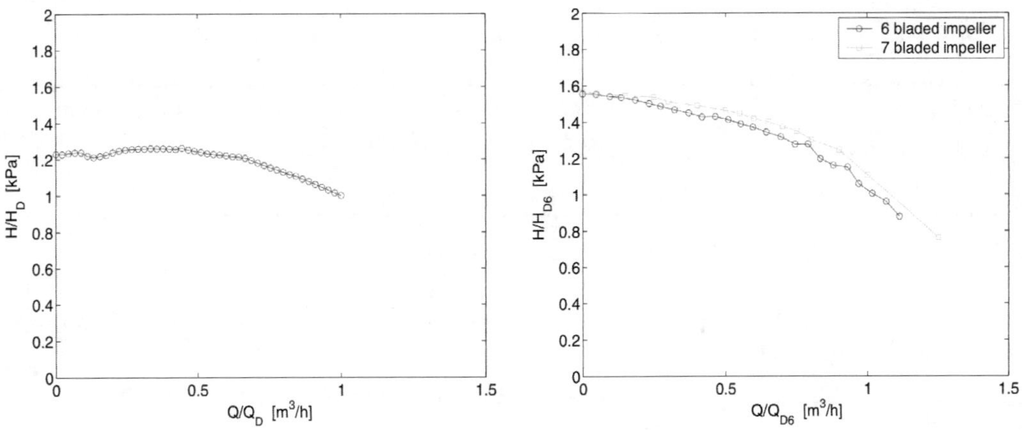

Figure 3. Performance curves : (i) CR90 pump (ii) CR4x2 pumps with 6 and 7 bladed impellers.

performance curves for CR4x2, for both 6 and 7 bladed impellers, are relatively smoother and less flat at low flow rates.

3. Proposed Methodologies for Detection of Stall

Two different methodologies were proposed for detection of stall by analysis of pressure and velocity time series acquired from the two pumps. First, the raw data were analysed directly by breaking up the time series into time-periodic strips, assuming constant angular velocity of the impeller, and aligning these strips by introducing a third coordinate to generate a carpet plot of velocity/pressure data. It may then be possible to identify some flow feature, which is a manifestation of rotating stall, as it propagates from one blade passage to the next. Flow reversal could be detected in the time series by monitoring the sign of the radial velocity.

In the second approach, motivated by the results reported in [4], spectral analysis of uniformly sampled pressure and non-uniformly sampled velocity data were to be carried out to determine if the stall frequency could be clearly identified. The stall propagation frequency could be as low as 5-10 % of the impeller speed as reported in [3]-[4] and hence the success of this method would depend upon accurate resolution of low-frequency content of the power spectra corresponding to velocity and pressure time series.

4. Results and Discussions

At the outset, an order-of-magnitude analysis was performed to estimate the stall propagation frequency for each of the two pumps. This required proposing a mechanism that enables the calculation of residence time for a stall cell in a blade passage prior to its propagation to the one adjacent to it. The stall-cell residence time was taken to be of the same order as the flow residence time i.e. the time it takes a fluid particle to traverse the extent of the impeller blade. With a radial velocity of the order of 10 m/s and a blade camber line of order 0.25 m, a residence time of 0.025 seconds is obtained which implies that in 1 second the stall cell would have propagated 40 impeller blades. This would correspond to a stall propagation frequency of about 6 Hz since each of the two pumps has approximately the same number of blades in the impeller(6/7 blades). The geometry and operating parameters for the two pumps are indicated in the table below.

An order-of-magnitude estimate of stall-propagation frequency is an important step in the detection of stall. It should be mentioned that data acquisition for the CR90 pump was much more extensive as compared to CR4x2 pump. Not only were pressure and velocity time series acquired for CR90 pump,

Table 1. Relevant Geometry and Operating Parameters for the Two Pumps

Pump Code	Design Flow Rate	Impeller Speed	Impeller Blades	Diffuser Vanes
CR90	90 m^3/h	50 Hz	7	11
CR4x2	10.8 m^3/h	12.1 Hz	6/7	4

whereas data acquired for the CR4x2 pump was limited to velocity time series, but many more off-design conditions were monitored as well.

4.1. Analysis of CR90 data

The first approach that involved generating a carpet plot of raw data, which would reflect its inherent periodicity, did not provide much useful information for the detection of stall. The pressure and velocity data for a turbulent flow has a continuous variation of time scales and it was not possible to discern any meaningful low-frequency pattern in the carpet plot which could have been interpreted as a manifestation of stall.

Subsequently, a flow reversal index was constructed for each revolution of the impeller which would indicate the fraction of data samples with negative radial velocity. Thus, an index of 0 would indicate that all sampled data have positive velocity whereas an index of 1 corresponds to all negative radial velocities for an impeller revolution. The flow reversal index for the CR90 pump is shown in Fig. 4 for 1000 impeller revolutions and for flow rates ranging from 15 m^3/h to 65 m^3/h. It is observed that at a flow rate of 25 m^3/h, quite a few impeller revolutions have a flow reversal index greater than 0.5 which would indicate a likelihood of stall occurrence. However, this method does not provide conclusive evidence of stall nor does it provide any means to estimate the stall propagation frequency.

In the second approach, power spectra of pressure and velocity time series were calculated and plotted as a function of frequency. The pressure data, which is uniformly sampled in time, was analysed using Matlab's power spectral density script. The velocity data, which is non-uniformly sampled in time, required a Matlab script [7] that is based on the technique proposed in [8].

The spectral analysis methodology for unevenly spaced data developed in [8] slightly modifies the conventional definition of power spectrum to retain the simple statistical behaviour of the evenly spaced case. It is shown that with the modification, power spectrum analysis and least-squares fitting of sine waves to the data are exactly equivalent. An important practical consequence is that the modified procedure is better suited for detecting a periodic signal in the presence of noise.

It was decided to initially examine the power spectra for design conditions (90 m^3/h) corresponding to the CR90 pump to identify the various frequency peaks that are revealed in the radial and tangential velocity data. The most pronounced peak in the two spectra in Fig. 5, which are shown up to 500 Hz, corresponds to impeller blade passage frequency(50 Hz x 7 blades) that is clearly observed at 350 Hz. The impeller frequency at 50 Hz and several of its super harmonics, particularly those at 100 Hz and 150 Hz, can also be identified in this figure.

To detect stall at off-design conditions, the power spectrum was plotted up to 50 Hz so that the low-frequency region can be examined in considerable detail. In Fig. 6, the power spectra for radial and tangential velocities are shown for a low flow rate of 15 m^3/h where rotating stall is expected to occur. A pronounced spike at 3 Hz can be clearly observed in both power spectra and several other significant spikes can also be seen in the neighbourhood of this frequency. The value obtained is in agreement with the order-of-magnitude estimate of the rotating stall frequency obtained earlier.

At a slightly larger flow rate of 25 m^3/h, where rotating stall is expected to persist, the power spectra for the two velocity components are shown in Fig. 7 and the pronounced spike appears shifted slightly lower at about 2 Hz.

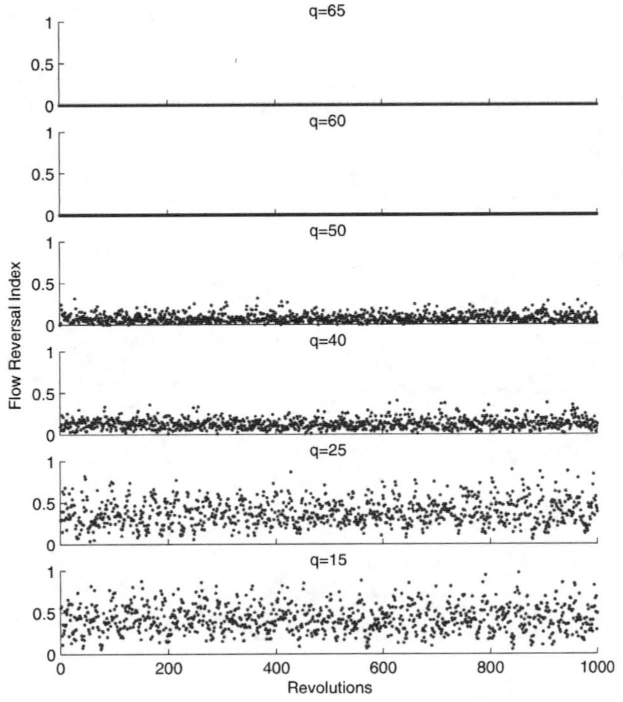

Figure 4. Flow reversal index based on the radial velocity for CR90 pump

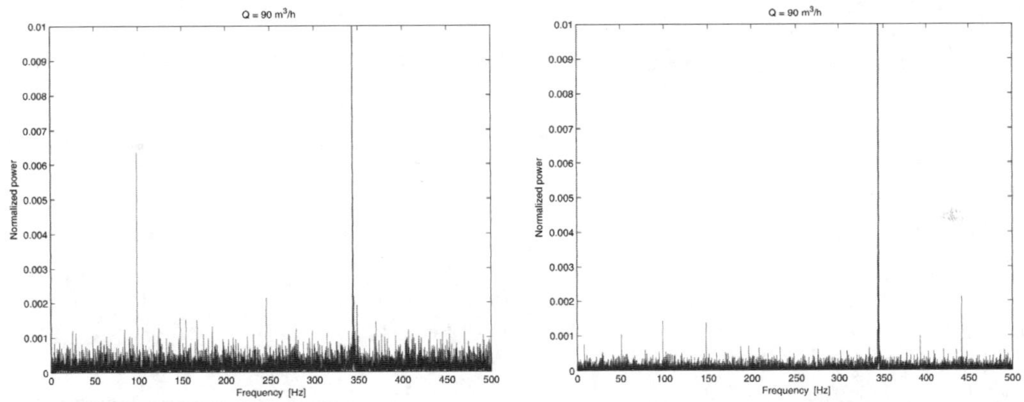

Figure 5. Power spectra at design conditions : (i) radial velocity (ii) tangential velocity

However, at a flow rate of 40 m^3/h, neither of the two velocity power spectra exhibit any noteworthy spike that can be interpreted as a manifestatation of rotating stall but the plots are not shown here to conserve space. It can then be conjectured that rotating stall appears somewhere in between a flow rate of 25 m^3/h and 40 m^3/h but there are no velocity measurements available at intermediate flow rates which can be used to further localize the onset of stall.

Pressure data, however, were acquired at intermediate flow rates for the CR90 pump. The power spectra for pressure data at flow rates of 25 m^3/h, 30 m^3/h, 35 m^3/h and 40 m^3/h are shown in Fig. 8, which clearly reveals a prominent spike at the lowest flow rate confirming the presence of rotating stall

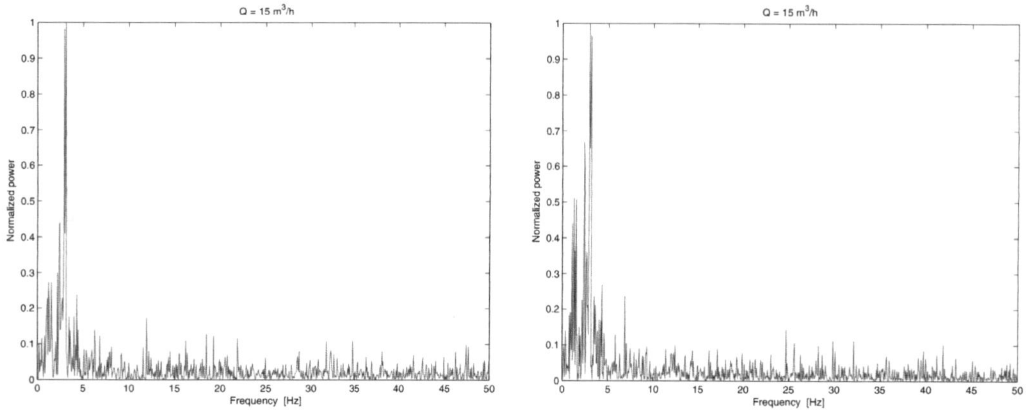

Figure 6. Off-design (15 m^3/h) power spectra : (i) radial velocity (ii) tangential velocity

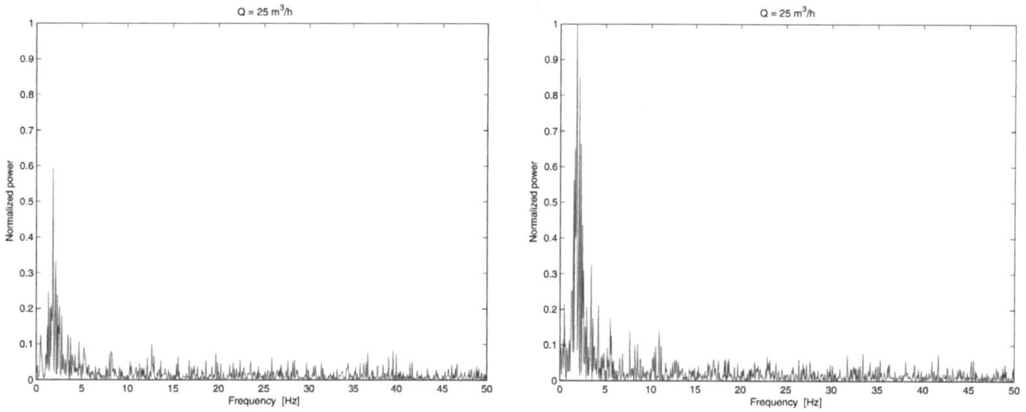

Figure 7. Off-Design (25 m^3/h) power spectra : (i) radial velocity (ii) tangential velocity

at a frequency of 2 Hz. It can also be inferred that rotating stall is absent at a flow rate of 40 m^3/h and to a less certain extent at 35 m^3/h as well. The power spectrum for a flow rate of 30 m^3/h reveals a large number of spikes which are not amenable to a straightforward interpretation; it is not clear if stall is present at this flow rate and whether it is stationary or rotating.

It would thus be prudent to conclude that the onset of rotating stall in the CR90 pump takes place somewhere in the neighbourhood of 30 m^3/h.

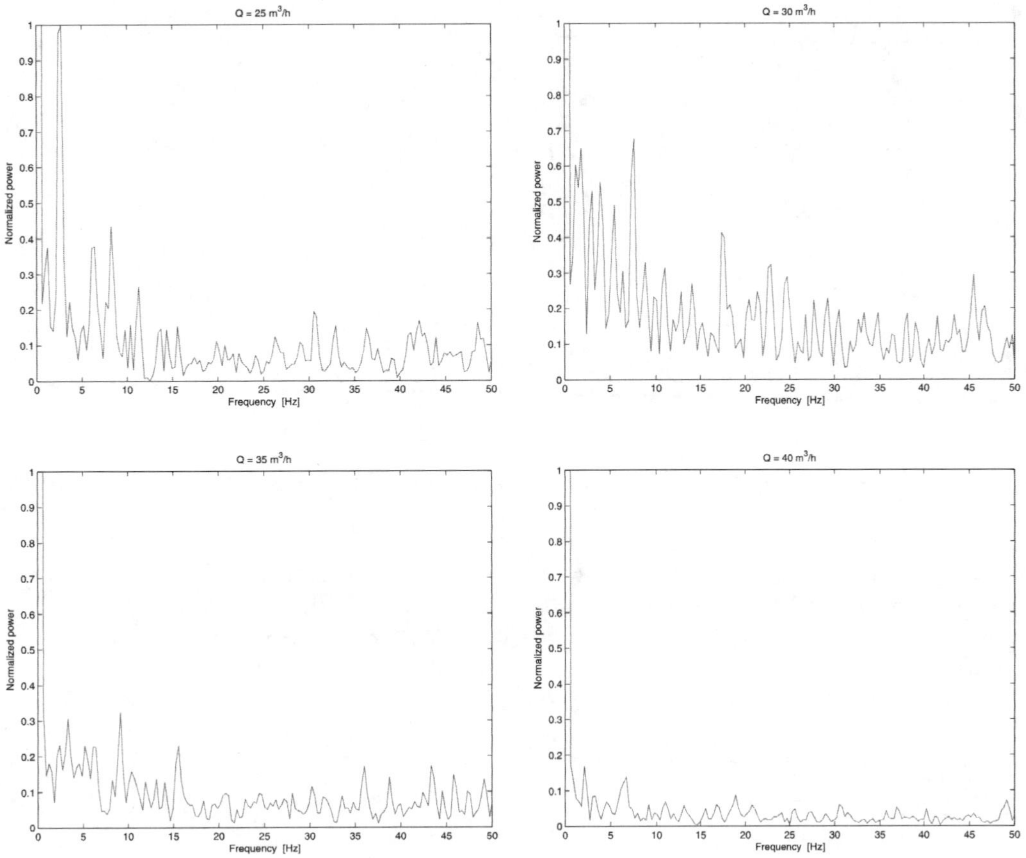

Figure 8. Pressure power spectra ; top left : 25 m^3/h, top right : 30 m^3/h, bottom left : 35 m^3/h, bottom right : 40 m^3/h

4.2. Analysis of CR4x2 data

Stall analysis, which was carried out in this study, for the CR4x2 pump was rather restricted in scope since limited data was acquired for this pump and the performance curve in Fig. 3 did not reveal any pronounced degradation at low flow rates for either of the two impellers. A stationary stall, however, was identified earlier for the 6 bladed impeller at an off-design flow rate of 2.7 m^3/h, and this was detected by an angular bin-averaging technique. It turns out that every alternate impeller blade passage gets stalled which is also reflected here in the power spectra for the tangential velocities in Fig. 9 corresponding to design and off-design conditions. Comparison of the two power spectra indicates that the impeller-blade passage frequency of 72.6 Hz (12.1 Hz x 6 blades) is reflected in both plots whereas a spike at half that value in the off-design case is due to stall cell periodicity.

Fig. 10 contains the tangential velocity power spectra for the 7 bladed impeller at design and off-design conditions. The impeller-blade passage frequency, which is now larger at 84.7 Hz (12.1 Hz x 7 blades), is clearly seen in both plots. However, the off-design case has several additional spikes and the plot appears noisier due to a factor of 4 magnification of the ordinate values relative to the design case. In the off-design spectra, the spikes at 39 Hz and 47 Hz are not subject to a straightforward interpretation even though the latter value is close to 48.4 Hz, which would correspond to an upstream influence of the diffuser vanes(12.1 Hz x 4 vanes). It is evident that a stationary stall cannot be sustained in the blade

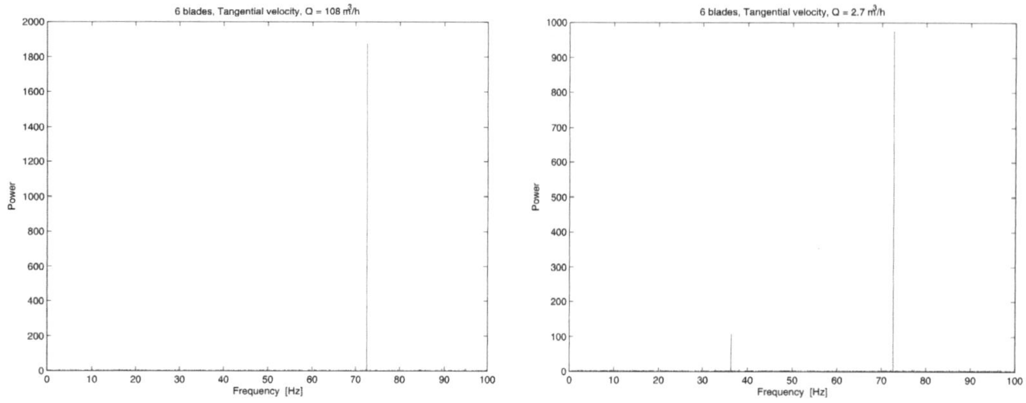

Figure 9. Tangential velocity power spectra (6 bladed impeller) : (i) 10.8 m^3/h (ii) 2.7 m^3/h

passages due to a loss of periodicity introduced by the odd number of impeller blades. The absence of a spike at frequencies below 12.1 Hz, which stands distinct from the background turbulent spectra, would largely rule out the existence of a rotating stall. It can then be stated that the analysis of time series for the 7 bladed configuration of the CR4x2 pump does not reveal any clearly identifiable stall frequency and this is also reflected in the smoother performance curve in Fig. 3 when compared to the 6 bladed case.

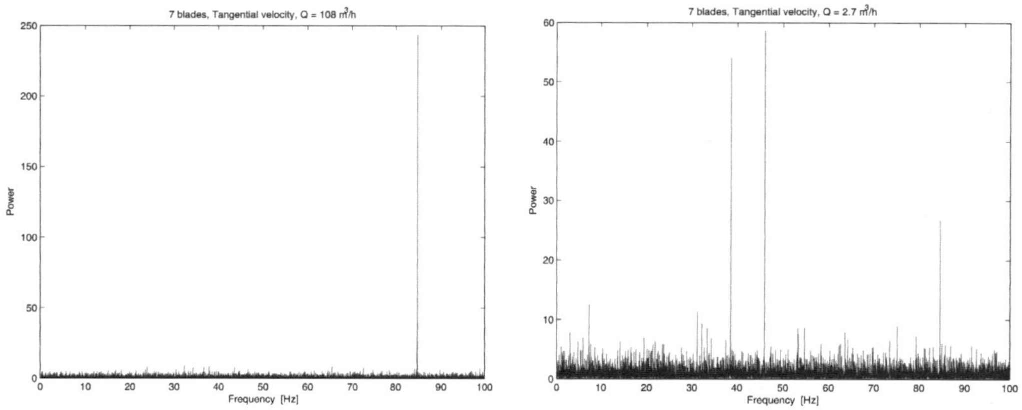

Figure 10. Tangential velocity power spectra (7 bladed impeller) : (i) 10.8 m^3/h (ii) 2.7 m^3/h

5. Conclusions

An analysis of rotating stall, based on experimental data provided by Grundfos A/S, was carried out. The data available for this study were pressure and velocity time series, acquired in or immediately downstream of the impeller in two different centrifugal pumps.

Two methodologies for detecting stall, based on the available experimental data, were proposed. In the first method, the time series were partitioned into strips based on the impeller revolution time and these strips were aligned to enable identification of a propagating pattern in the resulting carpet plot. However, no patterns were directly identifiable from the noisy plots due to the inherent fluctuations present in a time series for turbulent flow. A flow reversal index was generated, which would indicate the fraction of samples in each impeller revolution having negative radial velocity, that provided some qualitatively useful flow reversal information. However this method was limited in scope since it could neither directly detect stall nor a stall frequency.

The second method involved the analysis of power spectra corresponding to the velocity and pressure time series in order to determine if the stall propagation frequency, which is expected to occur in the low frequency part of the spectrum, could be isolated and accurately resolved. This method was successful in indicating if the pump was stalled or not and to some extent in identifying the frequency of rotating stall and the flow rate at which onset of stall occurs. However, not all of the data analysis for the two pumps yielded results which indicated the presence of stall. The velocity data from the CR90 pump were better conditioned and a flow rate of around 30 percent of the design value was identified as the location of stall onset. Furthermore, a stall frequency as low as 2 Hz, which is about 5 percent of the rotation frequency of the impeller, was found from the velocity power spectra. The power spectrum analysis for pressure data confirmed the results found in the velocity spectra. In the case of the CR4x2 pump, the occurrence of stationary stall phenomenon, which was detected by Pedersen et. al [5] for the 6-bladed impeller, was verified in this study from the power spectra. Nevertheless, there was no clearly identifiable stall frequency in the velocity spectra corresponding to the 7-bladed impeller for the CR4x2 pump.

Static pressure measurements, with transducers mounted flush with the wall, are significantly more convenient as compared to LDV measurements. Not only are sampling rates much higher for pressure measurements but more importantly this technique does not require a transparent test rig for data acquisition. The present study indicates that rotating stall frequency can be detected by spectral analysis of pressure time series just as well as velocity time series. However, more noise in the pressure spectra leads to less conclusive results and improvements may then be necesssary for future applications of this technique. The sampling frequency of 10 kHz might be too high for the scope of this work considering the frequency range of interest for stall detection. Stationary and rotating stall phenomena both occur at frequencies below the blade passage frequency and thus the sampling frequency can be adjusted accordingly using the Nyquist criterion. Furthermore, with some additional effort the utility of a dynamic pressure probe, located just outside the impeller, could be investigated as a device for stall detection. Spectral analysis of dynamic pressure could yield more prominent peaks and less noise which is very much desired at a flow rate corresponding to the onset of stall. However, this needs to be investigated further in order to characterize the performance of this intrusive measurement device for stationary and rotating stall detection.

Acknowledgments

This study was partially supported by Grundfos A/S, Bjerringbro, Denmark.

[1] Emmons H W, Kronauer R E and Rockett J A 1959 A survey of stall propagation-experiment and theory *ASME J. Basic Eng.* **81** 409-16

[2] Lennemann E and Howard J H G 1970 Unsteady flow phenomena in rotating centrifugal impeller passages *ASME J. Eng. Power* **92** 65-72

[3] Yoshida Y, Murakami Y, Tsurusaki T and Tsujimoto Y 1991 Rotating stall in centrifugal impeller/vaned diffuser systems *Proc. First ASME/JSME Joint Fluids Engineering Conference* **FED-107** 125-30

[4] Sinha M, Pinarbasi A and Katz J 2001 The flow structure during onset and developed states of rotating stall within a vaned diffuser of a centrifugal pump *ASME J. Fluids Eng. 123* 490-99

[5] Pedersen N, Larsen P S and Jacobsen C B 2003 Flow in a centrifugal pump impeller at design and off-design conditions-part I: Particle Image Velocimetry(PIV) and Laser Doppler Velocimetry(LDV) measurements *ASME J. Fluids Eng. 125* 61-72

[6] Byskov R K, Pedersen N and Jacobsen C B 2003 Flow in a centrifugal pump impeller at design and off-design conditions-part II: Large eddy simulations *ASME J. Fluids Eng. 125* 73-83

[7] http://wise-obs.tau.ac.il/~eran/MATLAB/Timeseries.html

[8] Scargle J D 1982 Studies in astronomical time series analysis. II. Statistical aspects of spectral analysis of unevenly spaced data *Ap. J. 263*, 835-853

Institute of Physics Publishing
doi:10.1088/1742-6596/52/1/005

Journal of Physics: Conference Series **52** (2006) 46–57
Mathematics for Industry in Denmark

Determination of Distance from a 2D Picture

J Gravesen[1,5], B Lassen[2], R Melnik[2,6], B Picasso[3], R Piché[4], N Radulovic[2] and L X Wang[2]

[1] Department of Mathematics, Technical University of Denmark, Matematiktorvet Building 303S, DK-2800 Kgs. Lyngby, Denmark
[2] Mads Clausen Institute, University of Southern Denmark, Grundvigs Allé 150, DK-6400 Sønderborg, Denmark
[3] Scuola Normale Superiore, Pisa, Italy
[4] Department of Mathematics, Tampere University of Technology, FIN-33101 Tampere, Finland

E-mail: `J.Gravesen@mat.dtu.dk`, `benny@mci.sdu.dk`, `rmelnik@wlu.ca`, `radulle@mci.sdu.dk`, `wanglinxiang@mci.sdu.dk`, `b.picasso@sns.it`, `robert.piche@tut.fi`

Abstract. An optical device is used to scan a cavity. In a single incident the scanner produces what can be considered a blurred image of the intersection curve between a plane and the cavity. Mathematically the image represents an intensity function and that is obtained by integrating a certain kernel along the intersection curve.

We suggest methods to determine the kernel and subsequently the intersection curve given the image. The methodology is tested with some success using an artificial, but realistic kernel and some synthetic images produced by this kernel.

Keywords: Optical distance measurement, inverse problem
MSC: 65K10, 68T45, 78A55

1. The problem

At the 47th Study Group with Industry in Gråsten, Denmark, Martin Valvik from Unisensor Inc. presented a problem concerning an optical scanner (Figure 1).

The scanner is meant to be moved inside a cavity and find its shape. The idea is first to determine a sequence of cross sections and then assemble these cross sections to give the full inner surface of the cavity. The plane which defines the cross section is fixed relative to the scanner so the position of the scanner needs to be known in order to place the cross section correctly in space. This is not the problem that concerns us here though. We only want to find the cross section relative to the scanner.

The scanner is (approximately) axisymmetric and light is transmitted in a ring round the axis down to a cone from where it is reflected radially out in a plane (approximately). If we imagine that the scanner is inserted into some cavity then the light from the scanner will hit (a part of) the sides of the cavity and form a "curve of light", the above mentioned cross section. The light will be reflected back from the sides of the cavity to the cone where it will be reflected upwards through a lens and finally make an image in the image plane.

[5] Corresponding author
[6] Present address: Wilfrid Laurier University, 75 University Avenue West, Waterloo, Ontario, N2L 3C5, Canada

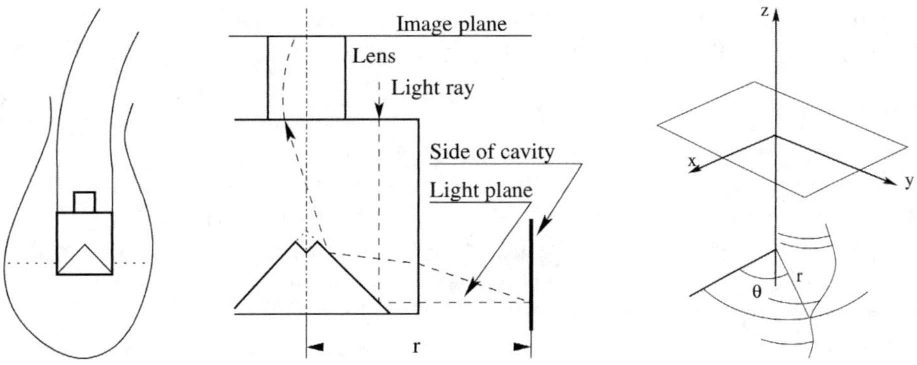

Figure 1. The scanner inside a cavity, the light path, and cylinder coordinates.

The problem is to determine the cross section from the image. So we know the intensity function $I(x, y)$ where (x, y) are coordinates in the image plane and we want to determine the function $r(\theta, z)$ where r is the distance from the axis of the scanner, θ is an angle around the center-axis (the z-axis), i.e., (r, θ, z) are cylinder coordinates.

We assume that the light is transmitted in a plane so the z dependence disappears and we are left with the task of determining the univariate function $r(\theta)$ from the intensity function $I(x, y) = I(\rho \cos\theta, \rho \sin\theta)$, where (ρ, θ) are polar coordinates in the image plane.

The paper is organized as follows. In Section 2 we explain the methodology presently used by the company and the assumptions the method is based upon and using the same assumptions we suggest an improvement to the method. in Section 3 we make a model based on more general assumptions. Mathematically we assume that the image is created by integration a certain kernel along the intersection curve. In Section 4 we explain how the kernel can be determined experimentally. We note that the problem at hand is related to shape-from-shading problems and we explore this connection further in Section 5. In this section we also explain how, by using the kernel mentioned above, we can obtain the intersection curve from an image by solving an optimization problem. In Section 6 we give an example of this procedure and the results are presented in Section 8. We did not have enough data to determine the kernel so instead we postulated a kernel we deemed reasonable. In Section 7 we use the singular value decomposition to check how well the postulated kernel could explain the image created by a point source. Finally in Section 9 we have the conclusions.

The people working on the problem was: Ole Brink-Kjær, Kasper Doring, Jens Graven, Jens Karlsson, Poul G. Hjorth, Benny Lassen, Roderick Melnik, Henrik Gordon Petersen, Bruno Picasso, Robert Piche, Janis Rimshans, Nenad Radulovic, Peter Røgen, Martin Valvik, Linxiang Wang, Stefan Wolff.

We will like take this opportunity to thank Unisensor and in particular Martin Valvik for contributing the problem, answering many questions during the study group, and even organizing new measurements when requested.

2. The present method and an improvement

The methodology presently used by the company builds upon a number of assumptions.

(i) It is assumed that the light from the scanner is reflected in a plane orthogonal to the axis so the only part of the cavity that can be "seen" is the intersection of the cavity with this plane, i.e, we only need to determine a function of one variable $r(\theta)$.

(ii) More severely, it is assumed that the image intensity along a radial ray, i.e., $I(x, y) = I(\rho \cos\theta, \rho \sin\theta)$ depends only on the distance $r = r(\theta)$ in this direction.

The point of view is that the point $\rho_{\max}$ where the function $I(\rho) = I(\rho\cos\theta, \rho\sin\theta)$ has its maximum is *the* image of the point $r(\theta)$, and that there is a one-to-one correspondence between r and $\rho_{\max}$. The function $\rho_{\max} \mapsto r$ can subsequently be determined experimentally. The scanner will never be exactly axisymmetric but that only means that $\rho_{\max}$ also depends on θ, i.e., we need to experimentally determine the function $(\theta, \rho_{\max}) \mapsto r$.

Given the assumptions 1) and 2) above, the basic problem is to determine the maximum intensity $I(\rho) = I(\rho\cos\theta, \rho\sin\theta)$ along a radial ray. First of all it is necessary to smooth the function before the maximum is determined, cf. Figure 4 where the lefthand graph shows the intensity along a radial ray. Furthermore the intensity function is often flat and hence it can be difficult and error prone to determine the maximum.

Instead of considering the maximum as the image of the point $r(\theta)$ it is better to consider the whole function $I(\rho) = I(\rho\cos\theta, \rho\sin\theta)$ as the result of a measurement and then use a more robust and easily calculated number to characterize the function. An obvious choice is to use the first moment (center of mass):

$$\rho_0 = \frac{\int_0^\infty \rho I(\rho)\,\mathrm{d}\rho}{\int_0^\infty I(\rho)\,\mathrm{d}\rho}, \tag{1}$$

as ρ_0 is insensitive to noise there is no need to perform any smoothing. The two functions $r(\rho_0)$ and $r(\rho_{\max})$ are both decreasing and exhibit the same qualitative behaviour. It is of course possible to incorporate asymmetry in this setting too, and determine the function $r(\theta, \rho_0)$ experimentally.

3. The basic model

We consider a cavity given by the equation $r = r(\theta, z)$ and assume that the light from the scanner is reflected diffusively with intensity $\alpha(r, z)$ from the point with cylinder coordinates $(r(\theta, z), \theta, z)$. If $(x, y) \mapsto G_0(x, y, r, \theta, z)$ is the intensity function created by a (unit) light source situated at the point (r, θ, z), then the intensity in the image plane created by the intensity function $\alpha(r, z)$ is found by integrating $G(x, y, r, \theta, z) = \alpha(r, z)G_0(x, y, r, \theta, z)$ over the surface of the cavity.

$$I(x, y) = \int_0^{2\pi} \int_{-\infty}^{\infty} G(x, y, r(\theta, z), \theta, z)\,\mathrm{d}z\,\mathrm{d}\theta. \tag{2}$$

A small remark is appropriate here. The intensity α at a point on the surface depends not only on the distance but also on the angle between the light ray and the normal of the surface. Also, the integration should be with respect to surface area and not $\mathrm{d}z\,\mathrm{d}\theta$. What happens is that light transmitted in small angle $\Delta\theta\Delta z$ can be distributed on a larger area of the surface and thus the intensity falls, but when we then integrate with respect to surface area we regain all the intensity, so the two corrections cancels each other, and (2) is correct.

In practice there would only be a small z-interval where $G(x, y, r(\theta, z), \theta, z) \neq 0$. If we can assume that the light from the scanner only emits light in a plane orthogonal to the symmetry axis, then we can drop the z dependence and we have diffuse light emitted from a curve with equation $r = r(\theta)$ and we have a kernel $G(x, y, r, \theta) = \alpha(r)G_0(x, y, r, \theta)$, where $\alpha(r) = 1/r$. The intensity is now

$$I(x, y) = \int_0^{2\pi} G(x, y, r(\theta), \theta)\,\mathrm{d}\theta, \tag{3}$$

and we want to determine the function $r(\theta)$. There are now two tasks

(i) Determine the kernel $G(x, y, r, \theta)$.

(ii) Provide a procedure that determines the function $r(\theta)$ from the intensity $I(x, y)$.

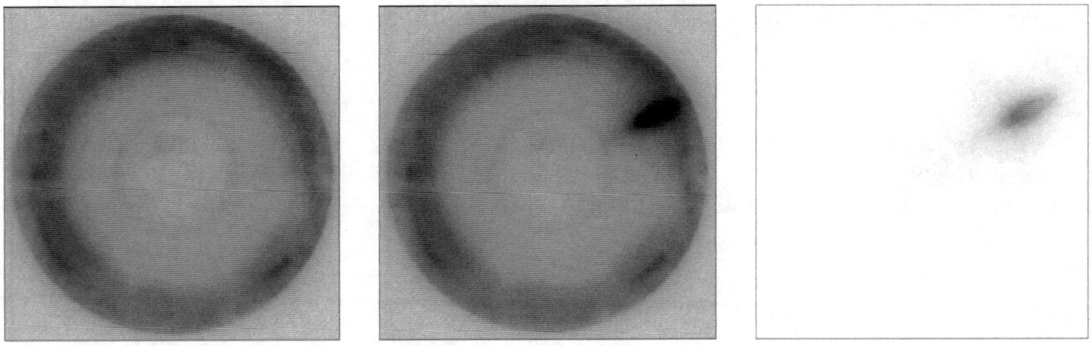

Figure 2. Inverted images: the reference, the thin stick, and the difference.

4. The kernel

If we assume that the scanner is symmetric then we have

$$G(x, y, r, \theta) = G(x \cos\theta - y \sin\theta, x \sin\theta + y \cos\theta, r, 0) \tag{4}$$

so we need only to determine $G(x, y, r, 0)$.

It is tempting to try to calculate the kernel from first principles. It is indeed possible to use geometric optics and follow a ray from a point in the "light"-plane through the scanner until it hits the image plane. This gives a map from a sphere centered at the light source to the image plane and the area deformation gives (a part of) the intensity, i.e., the kernel. The actually computations are rather involved and it will be hard (but not impossible) to get a usable result.

Unfortunately there is a more severe and fundamental problem. When the ray hits a boundary between two media, eg. at the cone or at the outside of scanner, a fraction of the light is transmitted and the other fraction gets reflected. These fractions depends not only on the angle between the ray and the surface normal, but also on the *polarization* of the light, cf. [7]. That in turn depends on the light source, the material of the cavity, and optical details of the scanner and would in practice be impossible to determine.

Instead we suggest to determine the kernel experimentally. The method would be to use the scanner, not to make a "picture" of a cavity, but to make a series of pictures of a thin stick in known distances r_k from the axis of the scanner. If the stick is sufficiently thin, then we can subtract a reference picture (a picture of nothing) and get a direct measurement of the kernel $G(x_i, y_j, r_k, 0)$. where (x_i, y_j) are the coordinates of the pixels of the CCD-camera, see Figure 2. Sufficiently thin would mean thinner than the final resolution that the scanner is supposed to provide.

After obtaining the values $G(x_i, y_j, r_k, 0)$ the next step is to approximate the kernel function with a suitable smooth function, for example a trivariate tensor product B-spline.

For a fixed r we can see from Figure 2 that the support of the kernel is concentrated in a relatively small area. Hence we concentrate the knots in that same area, cf. Figure 3. When we now let r vary two things happens. The shape of the intensity function changes, and the position of the support is shifted along the x-axis. We have two suggestions to model the kernel. The first is as a straight forward B-spline, cf. [3]

$$G(x, y, r, 0) \approx f(x, y, r) = \sum_{i,j,k} f^{i,j,k} N_i^{n_1}(x, \mathbf{x}) N_j^{n_2}(y, \mathbf{y}) N_k^{n_3}(r, \mathbf{r}), \tag{5}$$

where $N_i^n(t, \mathbf{t})$ denotes the ith basis function of degree n on the knot vector $\mathbf{t}$ and the knots $\mathbf{x}$ in the x-direction is chosen such that they can accommodate all positions of the support. The coefficients $f^{i,j,k}$ are chosen such that they minimize the 2-norm:

$$\sum_{i,j,k} \left(f(x_i, y_j, r_k) - G(x_i, y_j, r_k, 0)\right)^2 . \tag{6}$$

This is a linear problem which is easily solved. For the second suggestion we shift the origin to the first moment (the center of mass) of the intensity function, and propose an approximation to the kernel of the following form:

$$G(x, y, r, 0) \approx f(x - x_0(r), y, r) \tag{7}$$

where x_0 and f are B-splines

$$x_0(r) = \sum_i x_0^i N_i^{n_0}(r, \mathbf{r}_0), \tag{8}$$

$$f(x, y, r) = \sum_{i,j,k} f^{i,j,k} N_i^{n_1}(x, \mathbf{x}) N_j^{n_2}(y, \mathbf{y}) N_k^{n_3}(r, \mathbf{r}). \tag{9}$$

The coefficients x_0^i and $f^{i,j,k}$ are chosen such that they minimize the 2-norm:

$$\sum_k \left(x_0(r_k) - \frac{\sum_{i,j} x_i G(x_i, y_j, r_k, 0)}{\sum_{i,j} G(x_i, y_j, r_k, 0)} \right)^2 \tag{10}$$

and

$$\sum_{i,j,k} \left(f(x_i - x_0(r_k), y_j, r_k) - G(x_i, y_j, r_k, 0)\right)^2 \tag{11}$$

This leads to two linear problems which are easily solved. It will in both cases be necessary to experiment with the degrees (n_0,) n_1, n_2, n_3 and the number and placements of the knots ($\mathbf{r}_0$,) $\mathbf{x}, \mathbf{y}, \mathbf{r}$. The first suggestion needs more coefficients, but the structure is simpler so subsequent calculations are easier.

There is potentially another advantage of the form (5). Suppose the r-dependents of the kernel is so simple that it can be approximated by a low degree polynomial, of degree K. Then it is possible to pre-compute some values we will later need in a minimization procedure where we try to find the function $r(\theta)$ that makes $\int G(x_i, y_j, r(\theta), \theta)\, \mathrm{d}\theta$ fits the measurement best. Indeed, if $r(\theta)$ is a uniform B-spline, $r(\theta) = \sum_\ell a_l N_\ell^m(\theta - 2\pi\ell/n)$, then

$$\begin{aligned} G(x_i, y_j, r(\theta), \theta) &= f(x_i \cos\theta - y_j \sin\theta, x_i \sin\theta + y_j \cos\theta, r(\theta)) \\ &= \sum_{i,j,k} f^{i,j,k} N_i^{n_1}(x_i \cos\theta - y_j \sin\theta) N_j^{n_2}(x_i \sin\theta + y_j \cos\theta) \left(\sum_\ell a_\ell N_\ell^m \left(\theta - \frac{2\pi\ell}{n}\right)\right)^k . \end{aligned}$$

For $\theta \in [2\pi\ell/n, 2\pi(\ell+1)/n]$ only $m+1$ of the basis functions $N_\ell^m\left(\theta - \frac{2\pi\ell}{n}\right)$ are non-zero so we have

$$\int_0^{2\pi} G(x_i, y_j, r(\theta), \theta)\, \mathrm{d}\theta = \sum_{k=1}^{K} \sum_{\ell_1, \ldots, \ell_k} G_{i,j}^{\ell_1, \ell_2, \ldots, \ell_k} a_{\ell_1} a_{\ell_2} \ldots a_{\ell_k}, \tag{12}$$

where the inner sum is over sequences $\ell_1 \leq \ell_2 \leq \ldots \leq \ell_k$ with $l_k - l_1 \leq m$ and where the coefficients $G_{i,j}^{\ell_1, \ell_1, \ldots, \ell_k}$ can be precomputed. If the degrees m and K are sufficiently small then the number of coefficients $G_{i,j}^{\ell_1, \ell_1, \ldots, \ell_k}$ is manageable.

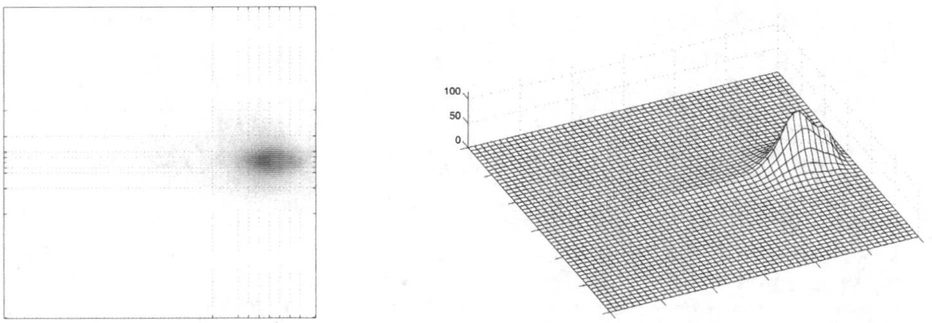

Figure 3. To the left the rotated image with the knots lines, to the right the quadratic B-spline approximating the intensity.

As a small test we have rotated the image in Figure 2, with 600 × 600 pixels, and values between 0 and 117. We have used a quadratic tensor product B-spline to approximate the intensity function, and the knot-vectors were

$$\mathbf{x} = 0, 400, 450, 470, 490, 510, 530, 550, 570, 600,$$
$$\mathbf{y} = 0, 200, 250, 280, 290, 300, 310, 320, 350, 400, 600.$$

The result can be seen in Figure 3. The approximation error measured in the max-norm, the 2-norm, and the 1-norm became

$$\|I - f\|_\infty = 15.6, \qquad \|I - f\|_2 = 1.27, \qquad \|I - f\|_1 = 0.50.$$

5. The procedure

We now have our approximation of the kernel $G(x, y, r, \theta)$, i.e., we assume that

$$G(x, y, r, \theta) = f(x\cos\theta - y\sin\theta, x\sin\theta + y\cos\theta, r), \qquad \text{or} \qquad (13)$$
$$G(x, y, r, \theta) = f((x - x_0(r))\cos\theta - y\sin\theta, (x - x_0(r))\sin\theta + y\cos\theta, r). \qquad (14)$$

Given measured intensities $I(x_i, y_j)$, we want to determine a function $r(\theta)$ such that (3) is satisfied. This is a non-trivial problem which is closely related to the so-called shape-from-shading problems. In what follows we highlight briefly this connection. Recall that our original problem described in Section 1 requires determining the shape of an object (a cavity in our case) from its two-dimensional images. In its essence, this is a special case of the classical shape-from-shading problem that has been discussed extensively in the literature [4, 10, 12, 9, 6, 13]. Basic steps upon which the model construction for this problem and the procedure for its solution can be based are as follows. As before, we will denote by $I(x, y)$ the intensity function of the image. In order to determine the shape (a surface) of the object on the basis of the given image(s), one could relate the intensity (or the brightness) function of the image (given, for example, at each pixel (x_i, y_j)) and the reflectance map that keeps the information on the real three-dimensional object. The form of the reflectance map, under which the images are generated from the (unknown) 3D depth map, has to be assumed as the only measurements produced by the scanner that can be analysed are the images themselves [13]. A natural way to relate such images to the 3D object is via the surface depth (the distance or the hight) function in the z-direction, denoted further by $u(x,y)$. Then, images of the same object, given in the (x,y)-plane, will differ from each other by shape gradients $\zeta_1 = \partial u/\partial x$ and $\zeta_2 = \partial u/\partial y$ which are essential characteristics of the 3D object. Hence, one can view the information based on the set of functions $u(x,y)$, ζ_1, and ζ_2 as the information from which, among other things, the shape of cavity can be extracted. As the cavity shape is modelled in (2) by using function $r(\theta, z)$ that enters the kernel of the integral equation, the development of specific algorithms to obtain r from (u, ζ_1, ζ_2) would be dependent on the specific form of kernel approximation in model (2). In order to determine u (and hence ζ_1 and ζ_2) recall that "shading", a 2D characteristic of the object, is represented mathematically by the image intensity $I(x,y)$, while "shape", a 3D characteristic of the object, is represented by the surface reflectance denoted here by $H(\zeta_1, \zeta_2)$. The latter specifies the reflectance of a surface under consideration as a function of its orientation and both functions are related to each other by the image irradiance equation (e.g., [4])

$$I(x,y) = H(\zeta_1(x,y), \zeta_2(x,y)), \quad \zeta_1 = \partial u/\partial x, \quad \zeta_2 = \partial u/\partial y. \tag{15}$$

Problem (15) can be reformulated as the first order partial differential equation under appropriate assumptions on function H. According to the Unisensor representative, the assumption of perfectly matt (Lambertian) surface is reasonable, hence we can use Lambert's law to determine a possible approximation to H. In particular, if a part of the surface with normal vector $\mathbf{n} = (-u_x, -u_y, 1)$ is illuminated in the direction $\mathbf{l} = (l_1, l_2, l_3)$, the emitted radiance (that will determine the image produced by the scanner) is given by the cosine of the angle between $\mathbf{n}$ and $\mathbf{l}$, namely by $(l_3 - l_1 u_x - l_2 u_y)/(u_x^2 + u_y^2 + 1)^{1/2}$. Therefore, if, for example, $\mathbf{l} = (0,0,1)$, then under the assumption of unit albedo and unit power of the source light, we arrive at the Hamilton-Jacobi equation of the form (e.g., [9]):

$$(u_x^2 + u_y^2 + 1)^{-1/2} = I(x,y). \tag{16}$$

In a more general situation, $\mathbf{l} = (\sin\sigma_l \cos\tau_l, \sin\sigma_l \sin\tau_l, \cos\sigma_l)$, where σ_l and τ_l are slant and tilt angles [2]. The latter expression can be simplified if we consider cross-sectional areas only, as depicted in Fig. 2, to $\mathbf{l} = (\cos\tau_l, \sin\tau_l, 0)$. Then, accounting for the surface albedo (denoted by λ) and the incident light flux (denoted by ρ), we obtained the following approximation to function $H(\zeta_1, \zeta_2)$:

$$H(\zeta_1, \zeta_2) = \lambda\rho \frac{-\zeta_1 \cos\tau_l - \zeta_2 \sin\tau_l}{\sqrt{1 + \zeta_1^2 + \zeta_2^2}}. \tag{17}$$

Given (15) and (17) we obtain the first order PDE with respect to the distance (or surface depth) function $u(x,y)$. According to the Unisensor representative, all parameters in this equation are measurable, including grayness of the surface λ, the incident light flux ρ (a characteristic of the scanner), while σ_l and τ_l can be estimated on the basis of the image intensity map (given by the company), physical characteristics of the surface and the scanner (e.g., [12, 13] and references therein). A natural way to solve the resulting problem is by applying the variational approach (e.g., [4, 8]). In this case, the cost functional can be defined as follows:

$$W = \iint_\Omega \left\{ [I - H]^2 + \mu_1 \left[\left(\frac{\partial^2 u}{\partial x^2}\right)^2 + 2\left(\frac{\partial^2 u}{\partial x \partial y}\right)^2 + \left(\frac{\partial^2 u}{\partial y^2}\right)^2 \right] + \mu_2 \Phi \right\} dx\, dy \to \min, \tag{18}$$

where the second term (known as the "departure-from-smoothness" error) is a regularizing term, as the problem at hand is an ill-posed problem (see, [4, 12, 5]). The first term (known as the "ring of light" or "brightness" error) comes into consideration directly from equation (15), while the third term is added to ensure integrability of the constrained optimization problem.

Models (2) and (17) are related in a sense that both are attempts to construct a model that relates the information about a 3D object and the information about 2D images of that object.

While in the former case, we use a geometric analysis of the problem to construct the model, the latter approach is ultimately based on a more conventional idea of using energy functionals (e.g., [1]). In our case, instead of using a more conventional approximation (17), we attempt to approximate the kernel of the reflectance map as it is stated by our formula (2). In this case, we model $r(\theta)$ (rather than function $u(x, y)$) as a uniform periodic B-spline

$$r(\theta) = \sum_{k=1}^{n} a_k N^m \left(\frac{n}{2\pi}\theta - k \right), \tag{19}$$

where N is the B-spline of degree m on the uniform knot sequence $\dots, -1, 0, 1, \dots$ and n is sufficiently large to give the resolution we want. The coefficients a_k are chosen such that the 2-norm

$$\sum_{i,j} \left(I(x_i, y_j) - \int_0^{2\pi} G(x_i, y_j, r(\theta), \theta)\, \mathrm{d}\theta \right)^2$$

is minimized. As the integral in (18) is replaced by a finite sum, the third term (18) is not needed. Furthermore, since the problem will be solved in terms of coefficients a_k which will determine $r(\theta)$ by (19), we use the quadratic term $\|a_k\|^2$ to regularize the problem. In its essence, our approach has some similarities with the Green's function approach to shape-from-shading (e.g., [11]). Now, the task at hand is the following optimisation problem:

$$\text{minimize} \sum_{i,j} \left(I(x_i, y_j) - \int_0^{2\pi} G(x_i, y_j, r(\theta), \theta)\, \mathrm{d}\theta \right)^2 + \mu \|a_k\|^2, \tag{20}$$

where μ is a parameter and $\|\cdot\|$ is a suitable seminorm, common choices include

$$\|a_k\|^2 = \sum_k a_k^2, \quad \|a_k\|^2 = \sum_k (a_k - a_{k-1})^2, \quad \|a_k\|^2 = \sum_k (a_k - 2a_{k-1} + a_{k-2})^2.$$

An important aspect of this type of problems is to obtain a good initial guess for $r(\theta)$. The scanner will be moved slowly inside the cavity so we would have a series of images, i.e., intensity functions, from which we want to determine planar sections of the cavity, i.e., $r(\theta)$. We expect the planar section of two subsequent measurements to be very similar, so the result of the optimization of one measurement, would be a very good initial guess for the next. To bootstrap this procedure and have a good guess for the very first image one could use prior knowledge of the cavity or the simplified procedure suggested in section 2.

In Section 6 we give an example of this procedure. We did not have sufficient data to make a reasonable B-spline model of the kernel so instead we postulate a kernel with an exponential decay and a $1/r$ dependence.

We conclude this section with the following remark. The problem at hand is a strongly non-linear problem and it would be advantageous to reduce it (at least for some special cases) to a sequence of linear problems. For model (17) it can be done if we use the assumption that the real shape of the object is smooth. Then, the equation $I(x, y) = H(\zeta_1, \zeta_2)$ can be approximated by using Taylor's expansion of the reflectance function (e.g., [10]). As a result, it is easy to find the Fourier transforms F_{ζ_1} and F_{ζ_2} in terms of the Fourier transform of of $u(x, y)$. This allows to relate the Fourier spectrum of the image and the Fourier transform of the surface. Finding Fourier's inverse of F_u will allow to estimate the object shape. This simplification should be useful when the Unisensor needs to analyse sufficiently smooth surfaces without substantial irregularities.

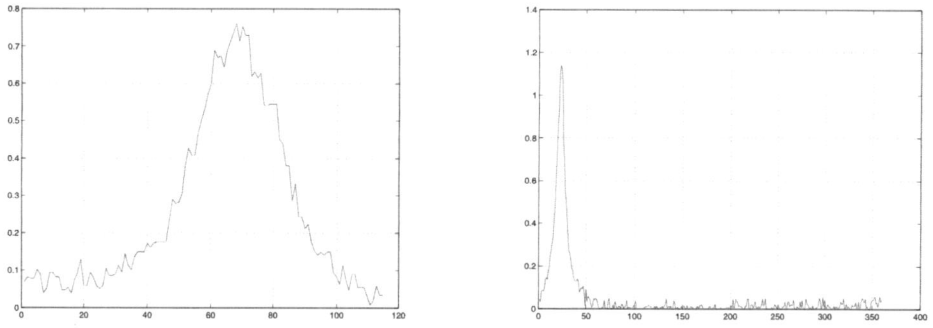

Figure 4. To the left the radial part $f(\rho)$ and to the right the angular part $g(\alpha)$ of intensity distribution $I(\rho,\alpha)$.

6. Case study example

For the case study we make some changes to the situation in Section 5. Instead of Cartesian coordinates (x, y) in the image plane we use polar coordinates (ρ, α), and instead of using B-splines to describe the kernel, we just postulate a kernel of the form

$$G(\rho, \alpha, r, \theta) = Ae^{-k_1(\rho - k_3/r)^2 - k_2(\alpha-\theta)^2} = Ae^{-k_1(\rho-k_3/r)^2}e^{-k_2(\alpha-\theta)^2}, \tag{21}$$

i.e., the product of two Gaussian. The parameters A, k_1, and k_2 are chosen such that the Gaussian approximates the the functions $f(\rho)$ and $g(\alpha)$ from Figure 4. The result was $A = 345$, $k_1 = 0.002$, and $k_2 = 0.01$. We had only data for three unknown values of r so we had no ways of estimating k_3 and we just put $k_3 = 1$. This gives us a predicted intensity as follows:

$$H = \int_0^{2\pi} 345e^{-0.002(\rho-1/r(\theta))^2 - 0.01(\alpha-\theta)^2}\,\mathrm{d}\theta. \tag{22}$$

The function r is represented by a vector $r = (r_1, r_2, ..., r_n)$, (can be thought as a discretisation in θ), and as regularization term we chose $\|r\|^2 = \frac{2\pi}{n}\sum_{i=1}^{n}(r_i)^2$. So the functional to minimise has the following form

$$\int_{\rho=0}^{\infty}\int_{\alpha=0}^{2\pi}\left\{[I-H]^2 + \mu\|r\|^2\right\}\,\mathrm{d}\alpha\,\mathrm{d}\rho \to \min. \tag{23}$$

where $I(\alpha, \rho)$ is the measured image intensity.

At this stage we took $r = 1$, corresponding to a unit circular object, and created a synthetic image using (22). We then tested the optimization procedure by letting the initial guess of r be randomly chosen around $r = 1$ with a relative error of 5%. The maximal initial relative error we have tried is 20%. Numerical experiments show that the optimization gives a quite good estimation of the actual value of r, as shown in figure 6 to 8.

7. Considerations regarding the Singular Values Decomposition

In this section we provide details of our SVD-based procedure to verify how well the kernel postulated in Section 6 can explain the image created by a point source. We have discretized ρ and α:

$$\begin{aligned} \rho &= \rho_i \qquad & i &= 1, \dots, N, \\ \alpha &= \alpha_j \qquad & j &= 1, \dots, M. \end{aligned}$$

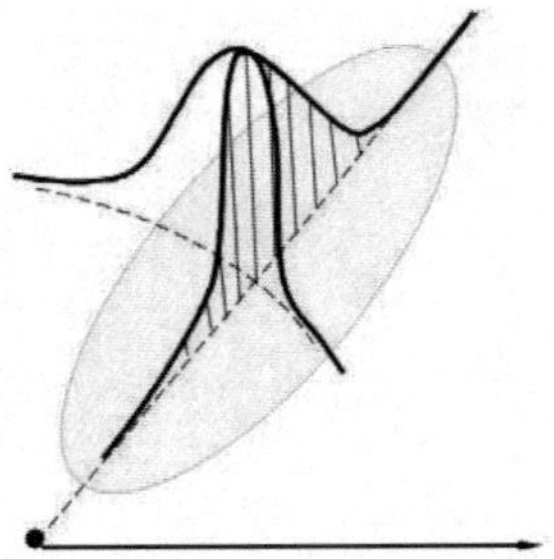

Figure 5. The assumption that $\mathrm{I}(\rho,\alpha)$ factors as is given by (24).

The intensity distribution $I(x,y)$ obtained processing the CCD picture generated by a source in position (r,θ) has been transformed into a discrete polar–coordinate function $\tilde{I}(\rho_i,\alpha_j)$: we denote by $\tilde{I}\in\mathbb{R}^{N\times M}$ the matrix $\tilde{I}_{i,j}\equiv\tilde{I}(\rho_i,\alpha_j)$.

The assumption (see Fig.5) that $I(\rho,\alpha)$ factors as

$$I(\rho,\alpha)=f(\rho)\cdot g(\alpha) \tag{24}$$

corresponds to the fact that the matrix $\tilde{I}$ has rank 1. Of course the numerical matrix $\tilde{I}$ has full rank, anyway we can get a measure of "how far" is that matrix from having rank 1 by analysing the singular values decomposition (SVD) of $\tilde{I}$. In fact:

$$\tilde{I}=USV^t \quad \text{with} \quad U\in\mathbb{R}^{N\times N}, \quad S\in\mathbb{R}^{N\times M}, \quad V\in\mathbb{R}^{M\times M},$$

where U and V are unitary matrices and $S=\mathrm{diag}(\sigma_i)$, $i=1,\dots,\min\{N,M\}$. The numbers $\sigma_1\geq\sigma_2\geq\dots\geq 0$ are the singular values of $\tilde{I}$. Denote by u_i and v_i the ith column of the matrices U and V respectively. If $\sigma_2\ll\sigma_1$ then $\tilde{I}$ is well approximated by the matrix $\tilde{I}_1=\sigma_1\cdot u_1v_1^t$ which has rank 1: it holds that $\|\tilde{I}-\tilde{I}_1\|_2\leq\sigma_2$. In this case $\sigma_1\cdot u_1$ and v_1 can be taken as a discrete version of $f(\rho)$ and $g(\alpha)$ respectively.

Indeed our calculations showed that there were at least 10 singular values not negligible: in that case the matrix is well approximated by $\tilde{I}_{10}=\sum_{i=k}^{10}\sigma_k u_k v_k^t$ that has rank 10 and is such that $\|\tilde{I}-\tilde{I}_{10}\|_2\leq\sigma_{11}$. If it is possible to find ten pairs of simple functions (f_k,g_k) such that $\sigma_k u_k\approx f_k(\rho_i)$ and $v_k\approx g_k(\alpha_i)$ then $\sum_k f_k(\rho)g_k(\alpha)$ is a good approximation to the kernel. This has not been investigated in further detail. Note also that for this procedure to be useful in practice, the r-dependence of the kernel should be simple.

8. Results of computations

To find optimal values for μ and n in (23) we chose to fix μ and then carry out calculations for $n=20,40$ and 80. We did calculations for $\mu=10^{-6},10^{-3},0.05$ and 0.1. In order to estimate the error in the minimized function we used the $\mathcal{L}^2$ norm on the difference between the exact function ($Z=1$) and the minimized function. Those numbers are shown in Table 1 together with the time it took to carry out the minimizations. The best value we found was $\mu=0.05$ and the more discretisation points the better. In Figure 6 to 8 the initial guess and the minimized function are shown for different values of μ.

The target function is extremely simple (a constant), as is the kernel, so it is not possible to draw any hard conclusions regarding the usuability of the procedure. That will require more realistic data and extensive testing.

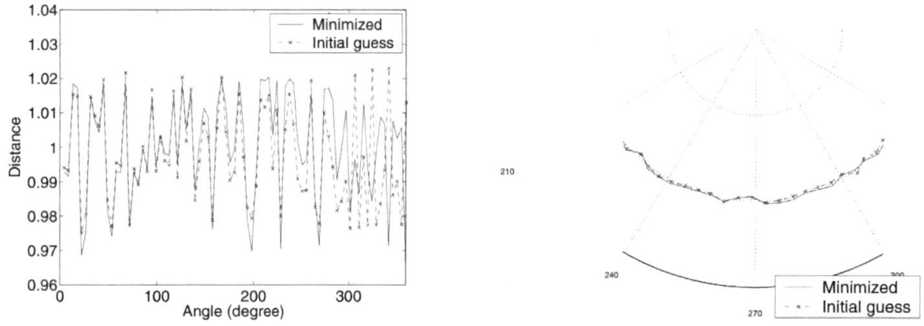

Figure 6. To the left a plot of the initial guess and the minimized distance function as a function of angle, to the right a section of a plot the initial guess and minimized function, for $\mu = 10^{-6}$ and $n = 80$.

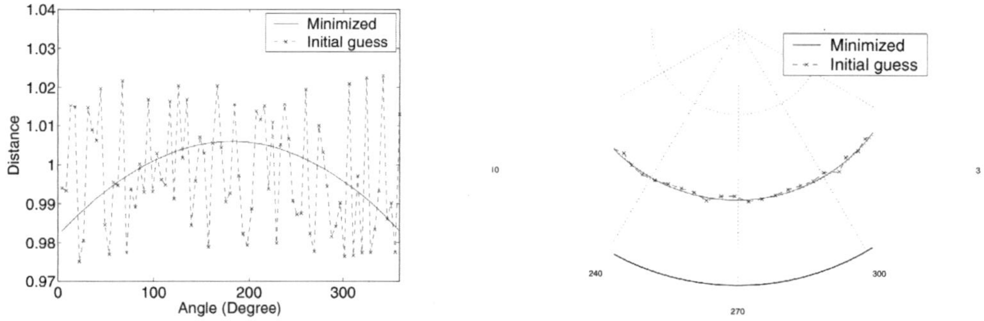

Figure 7. To the left a plot of the initial guess and the minimized distance function as a function of angle, to the right a section of a plot the initial guess and minimized function, for $\mu = 0.05$ and $n = 80$.

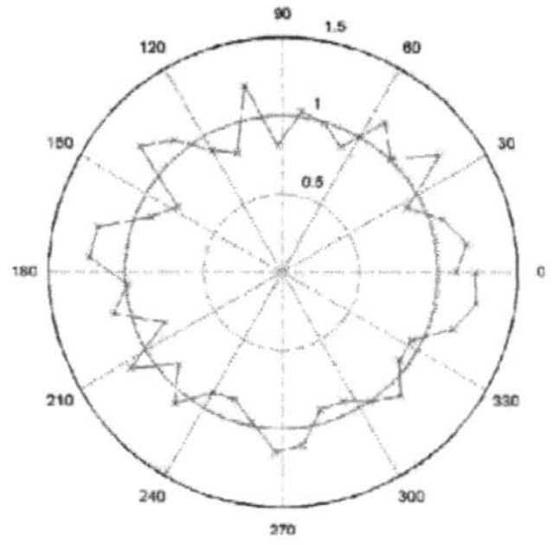

Figure 8. A plot of the the initial guess and minimized function for $\mu = 0.05$ and $n = 80$. The initial guess was generated with 20% random noise

9. Discussion and conclusion

We have demonstrated that the image produced by the scanner can be calculated by integrating a kernel along the "curve of light", cf. (3). The kernel can for a fixed r be represented well by a tensor product B-spline of a reasonable size, cf. (5) or (7).

$\mu \setminus n$	20		40		80	
10^{-1}	0.22%	17 sec	0.16%	61 sec	0.11%	464 sec
0.05	0.17%	12 sec	0.12%	52 sec	0.08%	90 sec
10^{-3}	0.29%	6 sec	0.15%	36 sec	0.09%	179 sec
10^{-6}	0.3%	5 sec	0.27%	6 sec	0.17%	80 sec

Table 1. The relative error and timing for different values of n and μ.

As an other approach, the singular value decomposition suggest that it is possible to describe the kernel as the sum of the products of ten pairs of univariate functions, cf. Section 7. So it might be possible to use something more simple than the tensor product B-spline descriptions, the univariate functions above would have to be described by a few parameters. But we have not investigated that.

We did not have enough data to model the r dependence of the kernel so there is a need for more work here, first of all images of "point sources" in known distances are needed.

Using a variational approach we have shown that it is possible to recapture the "curve of light" from the image. As the r dependence was unknown we used an artificial but realistic kernel, and we expect to get similar results using a more correct kernel. Once more it is necessary to do extensive testing with real data.

The timings given in Section 8 are not encouraging, but improvements can be made. The function $r(\theta)$ depends linearly on the coefficients a_k so it is not hard to get the gradient and Hessian of the objective function given in (20), all it takes is to differentiate (13) or (14) with respect to r, i.e., we only have to differentiate a number of B-splines. Access to the gradient and Hessian will speed the optimization up substantially, as will an implementation in C or Fortran. Also it is perhaps possible to pre-compute a substantial part of the objective function, cf. (12). Finally, since we deal with an inverse problem, some prior knowledge of the cavity should help improve the performance of the methodologies developed and tested in this contribution.

References

[1] Chan, T.F., Variational PDE models in image processing, *Notices of the AMS*, **50(1)**, 2003, 14–26.

[2] Chojnacki, W., Brooks, M.J., Gibbins, D., Revisiting Pentland's estimator of light source direction, *J. Opt. Soc. Am. A*, **11**, 1994, 118–124.

[3] de Boor, C., *A Practical Guide to Splines*, Springer-Verlag, New York 2001.

[4] Horn, B.K.P. and Brooks, M.J., *Shape from Shading*, MIT Press, Cambridge, MA, 1989.

[5] Hsieh, J.-W. et al Wavelet-based shape from shading, *Graphical Models and Image Processing*, **57(4)**, 1995, 343–362.

[6] Jones, A.G. and Taylor, C.J., Robust shape from shading, *Image and Vision Computing*, **12(7)**, 1994, 411–421.

[7] Landau, L.D. and Lifshitz, E.M., *Electrodynamics of Continuous Media*, Course of Theoretical Physics Vol. 8, Pergamon Press, Oxford 1960.

[8] Lepsky, O., Hu, C., and Shu, C-W., Analysis of the discontinuous Galerkin method for Hamilton-Jacobi equations, *Appl. Numerical Mathematics*, **33**, 2000, 423–434.

[9] Lions, P.L., Rouy, E., and Tourin, A., Shape-from-Shading, viscosity solutions and edges, *Numerische Mathematik*, **64(3)**, 1993, 323–353.

[10] Pentland, A.P., Linear shape from shading, *Int. J. of Computer Vision*, **4**, 1990, 153–162.

[11] Torreao, J. R. A., A Green's function approach to shape from shading, *Pattern Recognition*, **34**, 2001, 2367–2382.

[12] Zheng, Q. and Chellappa, R., Estimation of illuminant direction, albedo, and shape from shading, *IEEE Trans. on Pattern Analysis and Machine Intelligence*, **13(7)**, 1991, 680–702.

[13] Zhao, W. and Chellappa, R., Symmetric shape-from-shading using self-ratio image, *Int. J. of Computer Vision*, **45(1)**, 2001, 55–75.

Institute of Physics Publishing
doi:10.1088/1742-6596/52/1/006
Journal of Physics: Conference Series **52** (2006) 58–72
Mathematics for Industry in Denmark

Arrival-Time Detection and Ultrasonic Flow-Meter Applications

Morten Willatzen[1], Peter Søndergaard[2], Carl Latino[3], Frands Voss[1], Niels Lervad Andersen[1], Martin Brokate[4], Aicha Bounaim[5]

[1]University of Southern Denmark, Denmark
[2]Technical University of Denmark, Denmark
[3]Oklahoma State University, USA
[4]Technical University of Munich, Germany
[5]University of Oslo, Norway

E-mail: `willatzen@mci.sdu.dk`

Abstract. The Danfoss problem on ultrasonic flow measurement has been separated into three parts each handled by a subgroup of the authors listed above. The first subgroup deals with a presentation of modelling equations describing the physics of ultrasonic flow meters employing reciprocal ultrasonic transducer systems. The mathematical model presented allows the electrical output signal to be determined corresponding to any time-dependent electrical input signal. The transducers modelled consist of a piezoceramic material layer and a passive acoustic matching layer. The second subgroup analyzes the possibility of coding the input signal so as to simplify arrival-time detection by refinding the coded input sequence in the received signal. The narrow-band nature of the transducers makes this problem non-trivial but suggestions for improvement are proposed. The analysis given is based on traditional auto- and cross-correlation techniques. The third subgroup attempts to improve existing correlation methods in determining arrival-time detection of signals. A mathematical formulation of the problem is given and the application to a set of real signals provided by Danfoss A/S is performed with good results.

1. Introduction

An investigation of possible ways for the electronic excitation of ultrasonic transducer reciprocal systems is presented with emphasis to arrival-time detection accuracy for use in ultrasonic flow meters. This is done in three steps: (1) by analyzing the physics of ultrasonic reciprocal systems allowing for material-property differences between the various layers in the transducers [1, 2, 3] and the possibility to excite the system with a general input electrical signal (this includes the development of a MatLab reciprocal-transducer program), (2) by examining the influence of changing the input electrical signal using different coded signal inputs so as to detect more accurately and unambiguously the arrival time of the received signal based on auto- and cross-correlation techniques, and (3) propose new signal-processing methods and check them for consistency versus typical experimentally acquired ultrasonic signals.

[1] corresponding author

2. Theoretical Modelling of Reciprocal Transducer Systems

Figure 1 illustrates the transmission of an imposed voltage signal $u(t)$ across the transmitter electrodes to the received voltage signal $y(t)$ generated across the receiver electrodes. Henceforth, the transfer function components $H_1(t)$, $H_2(t)$, and $H_3(t)$ account for individual contributions to the total transfer function $H(t)$ from the transmitting transducer, the sound transmission between the transmitter and receiver, and the receiving transducer, respectively, i.e.,

$$H(t) = H_3(t)H_2(t)H_1(t), \tag{1}$$

satisfying

$$H(t) = \frac{y(t)}{u(t)}. \tag{2}$$

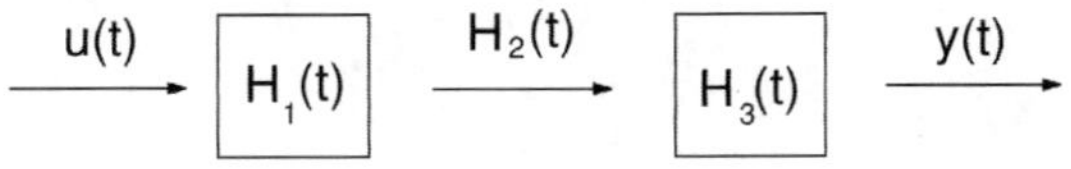

Figure 1. Schematic of the transfer function for the reciprocal transducer system setup.

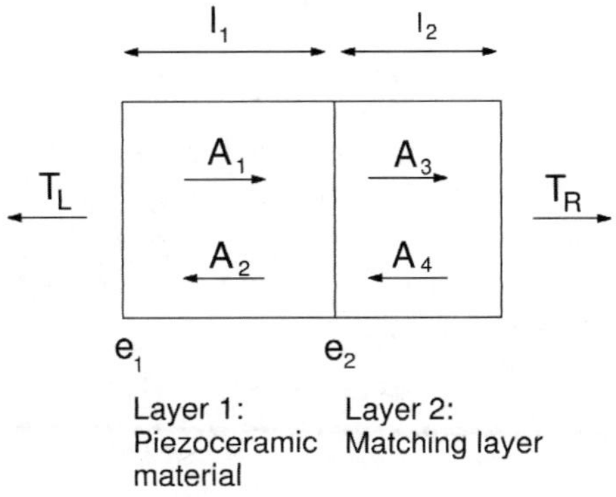

Figure 2. Schematic of the transducer with piezoceramic layer and matching layer.

Since the system modelled is a linear system, we can determine $y(t)$ from $u(t)$ as follows:

$$\begin{aligned} u(\omega) &= \frac{1}{\sqrt{2\pi}} \int_{-\infty}^{\infty} u(t)\exp(i\omega t)dt, \\ y(\omega) &= H_3(\omega)H_2(\omega)H_1(\omega)u(\omega) = H(\omega)u(\omega), \\ y(t) &= \frac{1}{\sqrt{2\pi}} \int_{-\infty}^{\infty} y(\omega)\exp(-i\omega t)d\omega. \end{aligned} \tag{3}$$

In the following, we shall disregard the influence of the sound transmission part: $H_2(\omega)$ as it merely represents a time delay between the output from the transmitter (aperture pressure) and the input to the receiver (incoming pressure wave) in the case of a vanishing fluid flow. In other words, we assume $H_2(\omega) = 1$. Next, we concentrate on determining $H_1(\omega)$ with the remark that $H_3(\omega)$ can be found following a similar procedure.

2.1. Determination of $H_1(\omega)$

A schematic figure of the transmitter is given in Figure 2. The governing differential equations are Newton's Second Law, the definition of strain, and Poisson's Equation:

$$\rho \frac{\partial u}{\partial t} = \frac{\partial T}{\partial z} \tag{4}$$

$$\frac{\partial u}{\partial z} = \frac{\partial S}{\partial t} \tag{5}$$

$$\frac{\partial D}{\partial z} = \rho_{free} = 0, \tag{6}$$

where ρ, ρ_{free}, u, T, S, D, t, and z are the mass density, free electric carrier density, particle velocity, stress, strain, electric displacement, time, and position, respectively. Employing the constitutive relation of a 1D piezoelectric material

$$T = -hD + c^D S, \tag{7}$$

where h and c^D are the piezoelectric h constant and bulk modulus, respectively, and Ohm's Law:

$$V(t) = Z_1 I + \int_{e1}^{e2} E(z,t)dz, \tag{8}$$

with $V(t)$, Z_1, I, $e1$, $e2$, and $E(z,t)$ the input voltage signal, the electric impedance connected in series with the transmitting transducer, the current, z-position of the left electrode, z-position of the right electrode, and the electric field, respectively, a complete set of equations needed so as to specify the dynamics of the transmitter circuit has been formulated.

Writing $V(t)$ in terms of its Fourier components:

$$V(t) = \frac{1}{\sqrt{2\pi}} \int_{-\infty}^{\infty} V(\omega) \exp(-i\omega t) d\omega \tag{9}$$

$$V(\omega) = \frac{1}{\sqrt{2\pi}} \int_{-\infty}^{\infty} V(t) \exp(i\omega t) dt, \tag{10}$$

and differentiating Equation (8) with respect to time allows a linear set of algebraic equations to be obtained for each frequency component $V(\omega)$:

$$T_L = c_1^D (A_2 + A_1) - h_1 D$$

$$\frac{T_L}{r_L} = c_1^D \left(\frac{A_2}{r_1} - \frac{A_1}{r_1} \right)$$

$$c_2^D (A_4 + A_3) = c_1^D \left(A_2 \exp(-ik_1 l_1) + A_1 \exp(ik_1 l_1) \right)$$

$$-h_1 D$$

$$c_2^D \left(\frac{A_4}{r_2} - \frac{A_3}{r_2} \right) = c_1^D \left(\frac{A_2 \exp(-ik_1 l_1)}{r_1} - \frac{A_1 \exp(ik_1 l_1)}{r_1} \right)$$

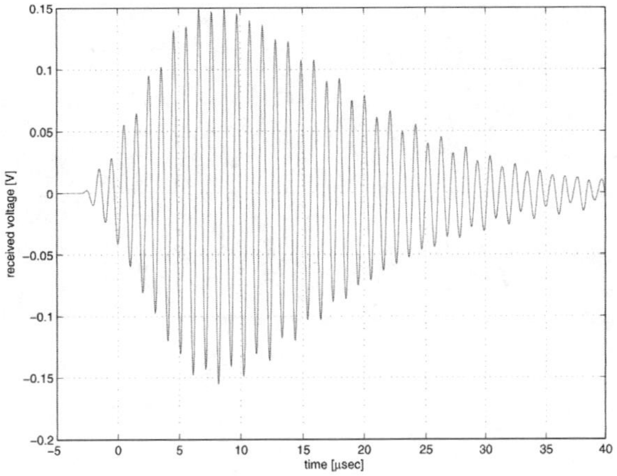

Figure 3. Output electrical signal for the case of a PZT5 piezoceramic transducer with steel matching layers (1 MHz frequency input, burst of eight sinus periods).

$$T_R = c_2^D\left(A_4\exp(-ik_2l_2) + A_3\exp(ik_2l_2)\right)$$
$$-\frac{T_R}{r_R} = c_2^D\left(\frac{A_4\exp(-ik_2l_2)}{r_2} - \frac{A_3\exp(ik_2l_2)}{r_2}\right)$$
$$-i\omega V(\omega) = -AZ_1\omega^2 D - i\omega\frac{l_1}{\epsilon_1^S}D - h_1\left[c_2^D\left(\frac{A_4}{r_2} - \frac{A_3}{r_2}\right) - c_1^D\left(\frac{A_2}{r_1} - \frac{A_1}{r_1}\right)\right], \quad (11)$$

where $A_i(i = 1, 2, 3, 4)$, $r_i(i = 1, 2)$, $T_L(T_R)$, $r_L(r_R)$, A, $l_i(i = 1, 2)$, and ϵ_1^S are the wave amplitudes for the four forward and backward propagating strain waves shown in Figure 2, the acoustic impedance of layer 1 and layer 2 materials, the stress wave amplitude at the left (right) aperture, the acoustic impedance of the left (right) approximately semi-infinite materials surrounding the transmitter, transducer cross-sectional area, the layer lengths of material 1 and 2, and the dielectric constant of the piezoelectric material at constant strain, respectively. In obtaining the set of Equations (11), use has been made of the relations:

$$I = A\frac{\partial D}{\partial t},$$
$$D = \epsilon^S E + \frac{h}{\epsilon^S}S, \quad (12)$$

where the expression for the current assumes that displacement currents are of main importance in an ultrasonic transducer operating at ultrasonic frequencies (close to 1 MHz).

The solution to Equation (11) determines T_R which (apart from a sign change) is the input pressure wave amplitude on the receiving transducer, i.e., $H_1(\omega)$ is found. A similar analysis of the receiver circuit allows the voltage component $V_{rec}(\omega)d\omega$ generated across the receiver electrodes to be calculated, i.e., $H_3(\omega)$ and $H(\omega)$ are found.

2.2. Case studies

In Figures 3 and 4, calculated output signals $y_1(t)$ and $y_2(t)$ are shown for the input voltage signal cases:

$$u_1(t) = \sin(\omega_0 t) \times H(4\mu\text{sec} > t > -4\mu\text{sec}), \quad (13)$$

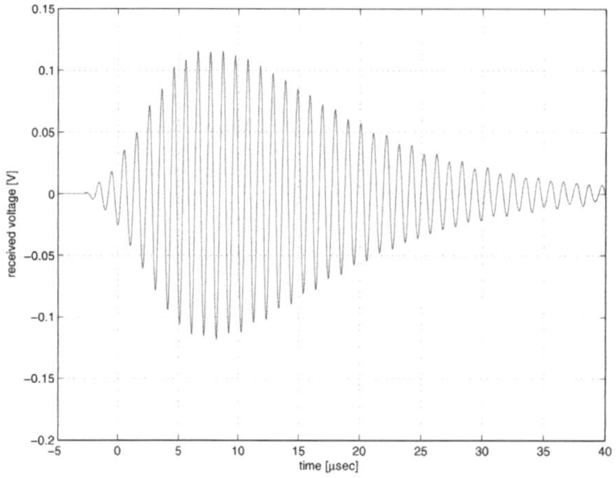

Figure 4. Output electrical signal for the case of a PZT5 piezoceramic transducer with steel matching layers (overlap of 50 kHz frequency input to the main 1 Mhz frequency input, burst of eight sinus periods).

and

$$u_2(t) = \sin(\omega_0 t) * \sin(\pi/2 + \omega_0 t/20) \times H(4\mu\text{sec} > t > -4\mu\text{sec}), \tag{14}$$

respectively. In Equations (13) and (14), $\frac{\omega_0}{2\pi} = 1$ MHz and

$$\begin{aligned} H(a < t < b) &= 1, \text{ if } a < t < b, \\ H(a < t < b) &= 0, \text{ elsewhere.} \end{aligned} \tag{15}$$

We finally note that the program delivered to Danfoss Flow Division can tackle *any* time-dependent voltage input $u(t)$.

3. Coded signals

Next, we present an investigation of the possibilities of coding the input signal to the transmitting transducer for improving identification of the transmitted signals and thus the time of flight.

The basis for this investigation is a simple electrical model of a pair of ultrasonic transducers modelled in MatLab [4, 5].

The following systems are modelled (refer to Figures 5-11):

1. Set of transducers with a centre frequency at 1 MHz

- a. Input signal is a single-frequency signal
- b. Input signal is a chirp signal
- c. Input signal is a two-frequency signal

All signals in the above cases 1a)-1c) are folded with an envelope (either a square- or a sine wave).

2. Set of transducers with different centre frequencies

- a. Transducer offset +/- 3 % (from 1 MHz)

- b. Transducer offset +/- 5 % (from 1 MHz)
- c. Transducer offset +/- 10 % (from 1 MHz)

All the above cases 2a)-2c) are investigated with dual frequency signals where the two frequencies are at the centres of the transducer as well as offset further to the 3dB crossing. All cases are for both types of envelopes. Furthermore, in case 2b), a chirp signal has been used and a few tests were made with noise added.

In Figure 5, a block diagram of the setup is shown.

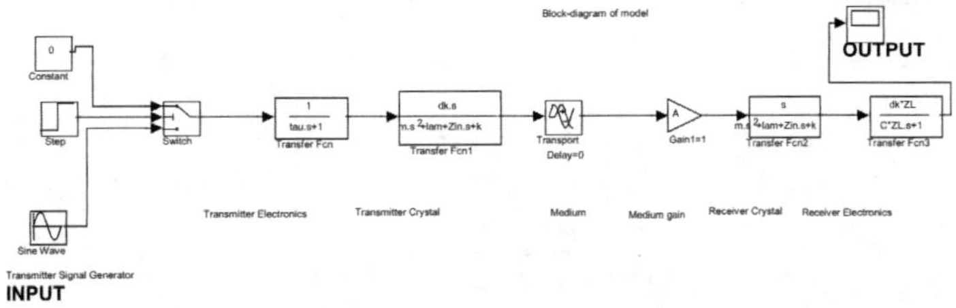

Figure 5. Block diagram in MatLab/Simulink of the setup.

The transfer function for the set of transducers having a centre frequency at 1 MHz is shown in Figure 6.

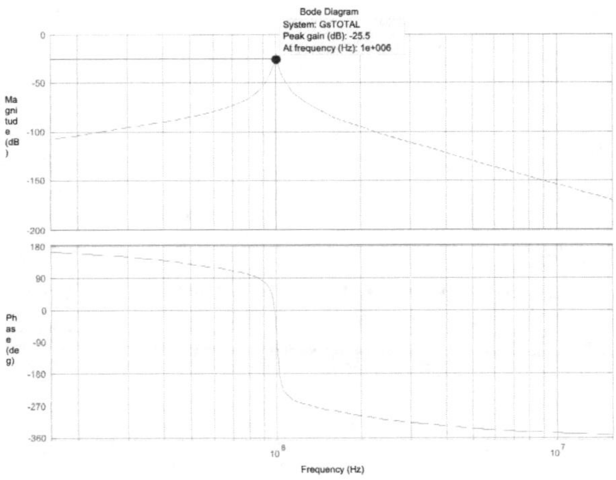

Figure 6. Bode diagram of the system transfer function. The upper curve is a plot of the magnitude as calculated in dB from −200 to 0 (y axis). The lower curve is a plot of the phase from −360 to +180 degrees (y axis). Both curves are plotted vs. frequency in the range 100 kHz to 10 Mhz.

Employing an enveloped input signal as shown in Figure 7(left) we obtain the output signal shown in Figure 7(right). The output signal is approximately of half the amplitude as compared to the output signal corresponding to a rectangular envelope.

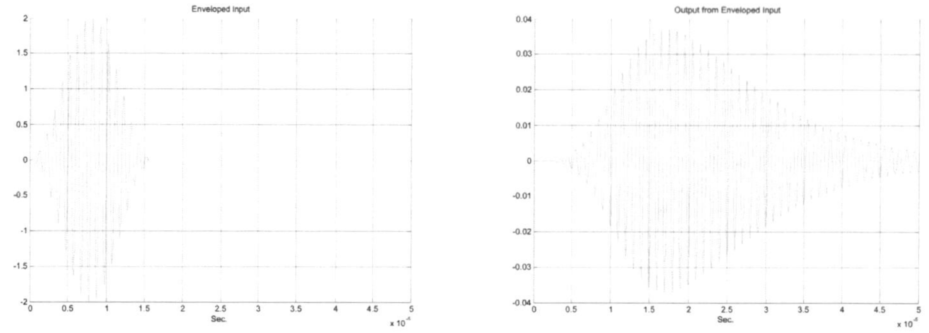

Figure 7. Left plot: Input electrical signal (in the range -2 V to $+2$ V) plotted vs. time in the range 0 to 50 microsecs. Right plot: Output electrical signal (in the range: -0.04 V to $+0.04$ V) plotted vs. time in the range 0 to 50 microsecs.

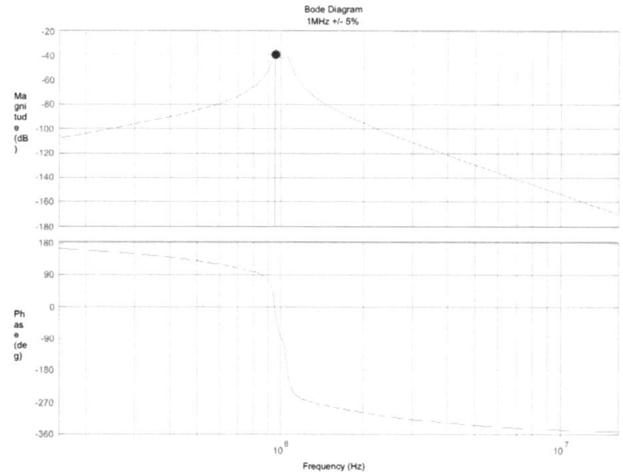

Figure 8. Bode diagram of the system transfer function. The upper curve is a plot of the magnitude as calculated in dB from -180 to -20 (y axis). The lower curve is a plot of the phase from -360 to $+180$ degrees (y axis). Both curves are plotted vs. frequency in the range 100 kHz to 10 Mhz.

Using a set of transducers with different resonance frequencies, say approximately +/- 5%, we get a transfer function as shown in Figure 8.

This combination is, of course, making the set of transducers slightly more broad-banded thus decreasing the signals moderately but - as a gain - increasing the frequency-band, allowing for a signal in the range +/- app. 8% relative to the original centre frequency to pass through the system.

Using an input signal consisting of two single frequencies, 0.96 and 1.04 MHz, respectively, convoluted with a sine envelope as shown in Figure 9, we obtain a fairly well-behaved output signal as shown in the upper right part of the same figure. The autocorrelation and cross-correlation shown in the lower part of Figure 9 is quite a bit more pronounced or sharper compared to the auto- and cross-correlation obtained with the 1 MHz single frequency system

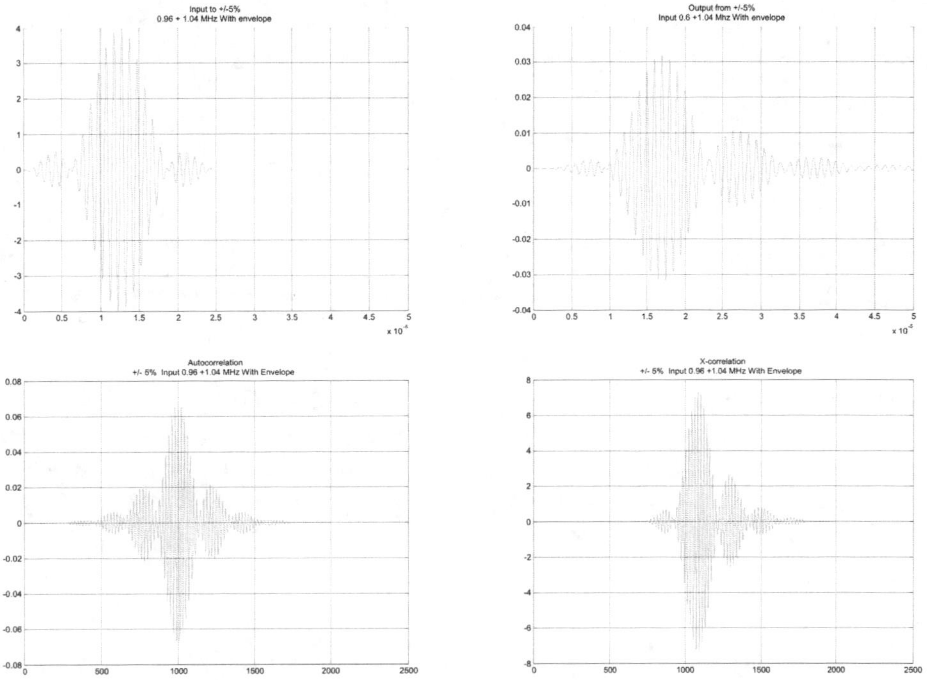

Figure 9. Upperleft plot: Input electrical signal (in the range -4 V to $+4$ V) plotted vs. time in the range 0 to 50 microsecs. Upperright plot: Output electrical signal (in the range: -0.04 V to $+0.04$ V) plotted vs. time in the range 0 to 50 microsecs. Lowerleft plot: Auto-correlation vs. sampling number (20 samples per microsecond; the y axis is from -0.08 to $+0.08$) and the x axis is from 0 to 2500 samples. Lowerright plot: Cross-correlation vs. sampling number (20 samples per microsecond). The y axis ranges from -8 to 8 and the x axis ranges from 0 to 2500 samples.

which, of course, is an improvement.

To investigate the noise immunity of the 'broadband system' 100% noise is added to the input signal. Superimposing the noise to a two-frequency input signal of 0.94 MHz and 1.06 MHz we get an output signal and a cross-correlation as shown in Figure 10. Even in this case the cross-correlation seems fairly pronounced while having a maximum peak easy to identify. Reducing the noise to a more moderate level of 10%, commonly seen in applications, and using a chirped signal varying from 0.94 MHz to 1.06 MHz we get an output signal and a cross-correlation as seen in Figure 11. Again a fairly sharp cross-correlation is obtained indicating that a coded transmitted signal could lead to improvements in the identification of received signal arrival times, although the amplitude is somewhat reduced.

4. Signal-processing methods for use in determining arrival times of ultrasonic signals

The third subgroup of the Danfoss Flow Division ultrasonic arrival-time group has examined possibilities for improving on the signal processing methods for detecting ultrasonic signals. It is necessary to introduce appropriate notation for the analysis described in this Section.

- All the calculations will be done exclusively in $\mathbb{C}^L$ using the dot-product and the Euclidean

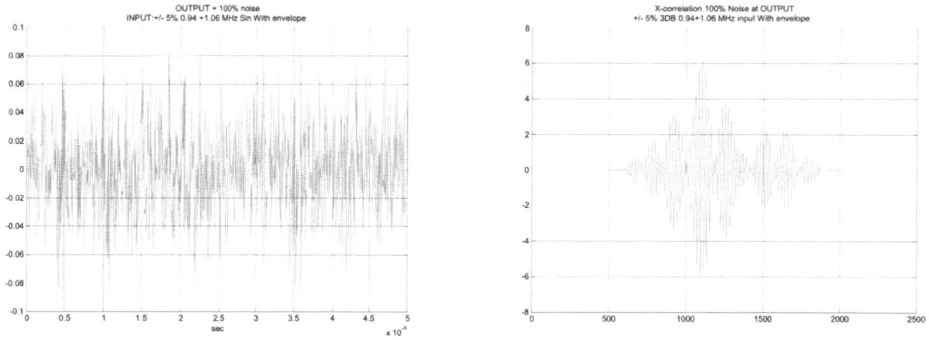

Figure 10. Left plot: Output electrical signal (in the range: -0.1 V to $+0.1$ V) plotted vs. time in the range 0 to 50 microsecs. Right plot: Cross-correlation vs. sampling number (20 samples per microsecond). The y axis ranges from -8 to $+8$ and the x axis ranges from 0 to 2500 samples.

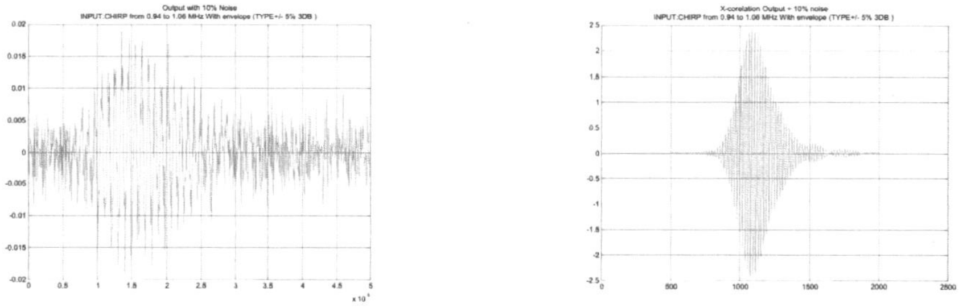

Figure 11. Left plot: Output electrical signal (in the range -0.02 V to $+0.02$ V) plotted vs. time in the range 0 to 50 microsecs. Right plot: Cross-correlation vs. sampling number (20 samples per microsecond). The y axis ranges from -2.5 to $+2.5$ and the x axis ranges from 0 to 2500 samples.

norm:

$$\|f\|_2 = \sqrt{\sum_{j=0}^{L-1} |f(j)|^2}, \tag{16}$$

where f is a signal of length L.

- For all $t = 0, ..., L-1$, the *translation operator* $\tau_t : \mathbb{R}^L \to \mathbb{R}^L$ (or $\mathbb{C}^L \to \mathbb{C}^L$) is defined by

$$(\tau_t f)(j) = f(j-t), \quad j = 0, ..., L-1. \tag{17}$$

Note that since the signals are of finite length, the translation is really a circular shift of the signal.
- For all $k = 0, ..., L-1$ the *modulation operator* $m_k : \mathbb{C}^L \to \mathbb{C}^L$ is defined by

$$(m_k f)(j) = e^{2\pi ikj/L} f(j), \quad j = 0, ..., L-1. \tag{18}$$

The *Discrete Fourier transform* $\mathcal{F} : \mathbb{C}^L \to \mathbb{C}^L$ is defined by

$$(\mathcal{F} f)(k) = \frac{1}{\sqrt{L}} \sum_{j=0}^{L-1} f(j) e^{-2\pi ikj/L}. \tag{19}$$

Because of this normalization:$\|\mathcal{F} f\|_2 = \|f\|_2$.
- The following relations hold for interchanging the operators:

$$\mathcal{F} \tau_t f = m_{-t} \mathcal{F} f \tag{20}$$

and

$$m_k \tau_t f = e^{2\pi ikt/L} \tau_t m_k f \tag{21}$$

4.1. The problem
The approach is to consider existing real-world data and to try to improve existing correlation methods. We present essentially a mathematical formulation of the problem and the application to some provided real signals. First, two real signals of length L are given. The upstream signal is denoted by $y_1 \in \mathbb{R}^L$ and the downstream signal by $y_2 \in \mathbb{R}^L$. Next, we propose to model the main difference between y_1 and y_2 as a time shift Δt, or equivalently, $y_1 \approx \tau_{\Delta t} y_2$.

The problem with this model is that the signals may be noisy, and the tails of the signal do not match very well. For illustration, some real world signals can be seen in Figure 12.

For each example, two signals are displayed. The first signal is the one which traveled *against* the flow direction (labelled "upstream") and the second signal is the signal which traveled *along* the flow direction (labelled "downsteam"). The signals correspond to a location after passage through the tube.

The signals are not very noisy and they consist mainly of one frequency close to 1 MHz. The differences between the up- and downstream signals are clearly visible.
The solution we propose is to try to match the signals in an appropriate norm that focuses on the part of the signals which are clearly identifiable.
In addition, for a correct operation of the flow meter, the time difference Δt must be determined with very high precision. In other words, the error on Δt must be less than the sampling distance. In this study, we have not tried to obtain such a high precision. However, we have managed to match the signals according to a precision within one wavelength (which for these examples is close to 20 samples). When Δt is determined with a better precision than one wavelength, the precision can be improved by studying the zero-crossings of the signal.

4.2. Time-frequency analysis
The *Discrete Short Time Fourier Transform* (DSTFT) with window $g \in \mathbb{C}^L$ is given by

$$: \mathbb{C}^L \to \mathbb{C}^{L \times L}, \; (V_g f)(m, n) = f \cdot m_m \tau_n g. \tag{22}$$

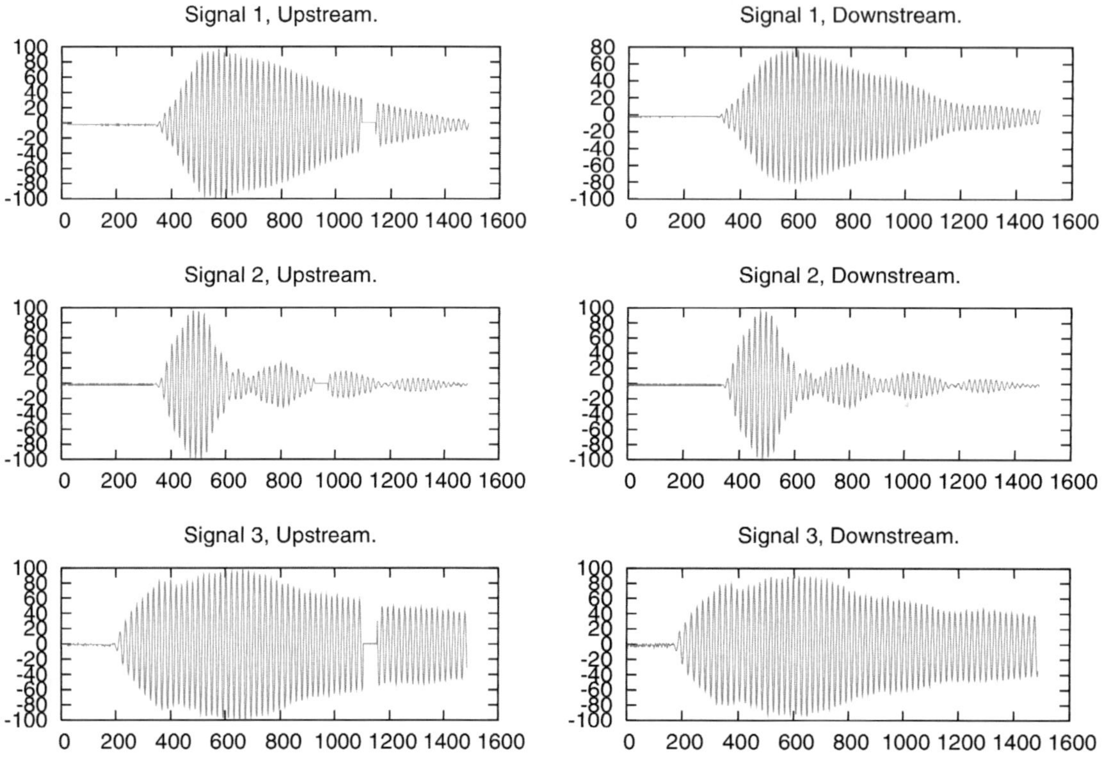

Figure 12. The examples are given to us by Danfoss Flow Division. Signals are plotted in arbitrary units vs. sampling number (20 samples per microsec).

Since the DSTFT maps $\mathbb{C}^L$ into $\mathbb{C}^{L\times L}$, it is a highly redundant transform. To avoid some of this redundancy, we are restricted to only the coefficients situated on a regularly spaced grid, i.e., to consider $c \in \mathbb{C}^{M\times N}$ given by

$$c(m,n) = f \cdot m_{mb}\tau_{na}g, \tag{23}$$

for some integers $a, b \in \mathbb{N}$ for which $L = Ma = Nb$ where $M, N \in \mathbb{N}$.
Depending on the parameters a, b and the window g it might be possible to reconstruct the original window from the coefficients c. This is studied in the theory of *Gabor frames*, see e.g., [8, 7].

4.3. Upstream-downstream flows time shift versus l^2-norms
In this subsection, we are interested in evaluating the time shift Δt between the upstream and the downstream flows as given above with respect to different norms. Δt is then calculated by minimizing, given the norm chosen, the difference between two signal data obtained for the two flows. The following subsections list the minimized quantities for five norms.

4.4. The standard l^2-norm
The first choice is to use the standard Euclidean norm. The optimization problem to solve becomes:

$$\Delta t = \arg\min_t \|y_1 - \tau_t y_2\|_2$$

$$= \arg\min_t \sum_{j=0}^{L-1} |y_1(j) - y_2(j-t)|^2, \quad (24)$$

or, equivalently, using the DFT

$$\begin{aligned} \Delta t &= \arg\min_t \|\mathcal{F}y_1 - m_{-t}\mathcal{F}y_2\|_2 \\ &= \arg\min_t \sum_{j=0}^{L-1} \left| (\mathcal{F}y_1)(j) - e^{-2\pi ijt/L} (\mathcal{F}y_2)(j) \right|^2. \end{aligned} \quad (25)$$

4.5. Weighted l^2-norm

Since the tails of the signals often differ, we propose a weighted l^2-norm, where $w \in \mathbb{R}^L$ is a weight-function that picks out the front of the signal by some heuristic approach. The optimization problem becomes:

$$\Delta t = \arg\min_t \sum_{j=0}^{L-1} w(j) |y_1(j) - y_2(j-t)|^2. \quad (26)$$

In order to roughly determine were the signal starts, we used the following primitive heuristic: Find the point in time where the signal first time goes above 50% of the maximum amplitude, and subtract four full wavelengths from this. This approach worked well for the few examples we considered. The important point to note is that while it is very difficult to determine the exact starting point of the signal, it is far easier to determine it plus/minus a few wavelengths.

4.6. The l^2-norm of a noise-filtered signal

The signals are very well localized in frequency around 1 MHz. In order to remove the noise, the standard technique of filtering the DFT can be used. The filtered signals $\tilde{y}_1$ and $\tilde{y}_2$ are determined by

$$(\mathcal{F}\tilde{y}_1)(\omega) = \alpha(\omega) (\mathcal{F}y_1)(\omega), \quad (27)$$

and similarly for y_2. Here α is a weight function that picks out the sought frequencies. The optimization problem becomes

$$\Delta t = \arg\min_t \sum_{j=0}^{L-1} |\tilde{y}_1(j) - \tilde{y}_2(j-t)|^2, \quad (28)$$

or equivalently,

$$\Delta t = \arg\min_t \sum_{k=0}^{L-1} \left| (\alpha\mathcal{F}y_1)(k) - e^{-2\pi ikt/L} (\alpha\mathcal{F}y_2)(k) \right|^2. \quad (29)$$

4.7. Weighted l^2-norm of noise-filtered signal

By combining the two previous methods, we obtain a joint time and frequency-filtering method. The corresponding optimization problem becomes

$$\Delta t = \arg\min_t \sum_{j=0}^{L-1} w(j) |\tilde{y}_1(j) - \tilde{y}_2(j-t)|^2, \quad (30)$$

where α, w, $\tilde{y}_1$, and $\tilde{y}_2$ are defined as previously.

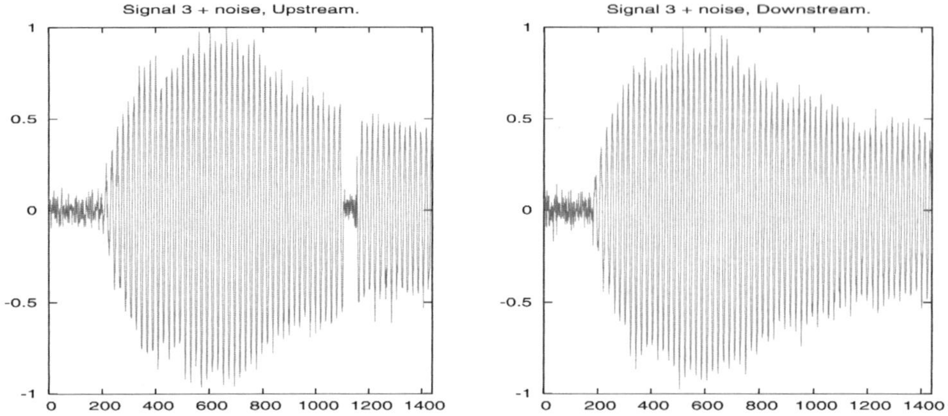

Figure 13. "Signal 3" with added noise. Signals are plotted in arbitrary units vs. sampling number (20 samples per microsec).

4.8. The l^2-norm of a selected part of the DSTFT

Alternatively, one can choose to consider only a cut-out of a sub-sampling of the DSTFT. The optimization problem to be solved is

$$\Delta t = \arg\min_t \sum_{m=m_1}^{m_2} \sum_{n=n_1}^{n_2} |y_1 \cdot m_{mb}\tau_{na}g - \tau_t y_2 \cdot m_{mb}\tau_{na}g|^2, \tag{31}$$

where the integers m_1, m_2, n_1, and n_2 are chosen by some heuristic approach such that they pick out the most informative part of the time frequency plane.

The method simply cuts out a rectangular part of the time-frequency plane. The place to cut is determined by the same heuristic approach as in the previous methods.

Using Equation (21),

$$\tau_{t+pa}y_2 \cdot m_{mb}\tau_{na}g \quad = \quad e^{-pamb/L}\tau_t y_2 \cdot m_{mb}\tau_{a(n-p)}g. \tag{32}$$

This shows that it is not necessary to compute $c_t(m,n) = \tau_t y_2 \cdot m_{mb}\tau_{na}g$ for all t, but only so for $t = 0, ..., a-1$, and then use that $c_{t+pa}(m,n) = e^{-pamb/L}c_t(m, n-p)$ so as to compute the remaining coefficients. In Ref. [9] it is shown how to compute $y_1 \cdot m_{mb}\tau_{na}g$ for all $m = 0, ..., M-1$ and $n = 0, ..., N-1$ using $\mathcal{O}(\frac{LN}{d})$ flops, where d is the greatest common denominator of b and N.

4.9. An example

As an application of the presented method, the example used is "Signal 3" from Figure 12 with added noise. The noise-filled signals can be seen in Figure 13.

Figure 14 shows how the time filtering and frequency filtering works.

The noise reduction removes the noise efficiently, but it also changes the envelope of the signal slightly. The time-filtering picks out the front of the signal by some degree of success.

The results for the different norms are shown on Figure 15. The signal appears almost identical to itself when shifted by a single wavelength and the fine details are smeared out due to noise. When no filtering is applied, the consequence is that the correct solution is hardly visible. Time filtering improves the case, and time-frequency filtering even more so. However, the method based on the DSTFT does not show an improvement over the simple time-frequency method.

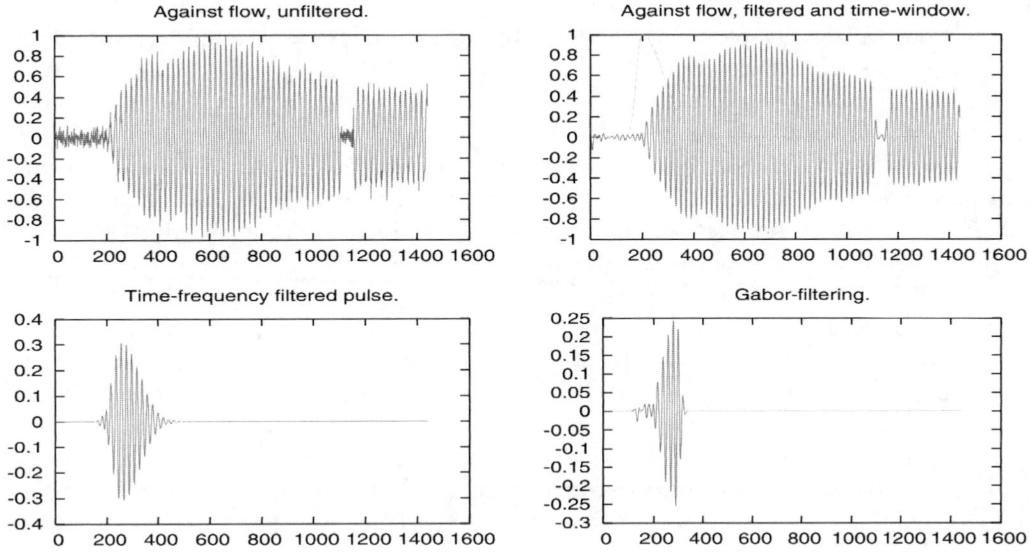

Figure 14. The figure shows how the variou6 filtering methods works. The first plot shows the original signal, the second plot shows the signal after noise reduction overlaid with the window to use for the time filtering. The third and fourth plot show how the signals look after the two different ways of time/frequency filtering have been applied. Signals are plotted in arbitrary units vs. sampling number (20 samples per microsec).

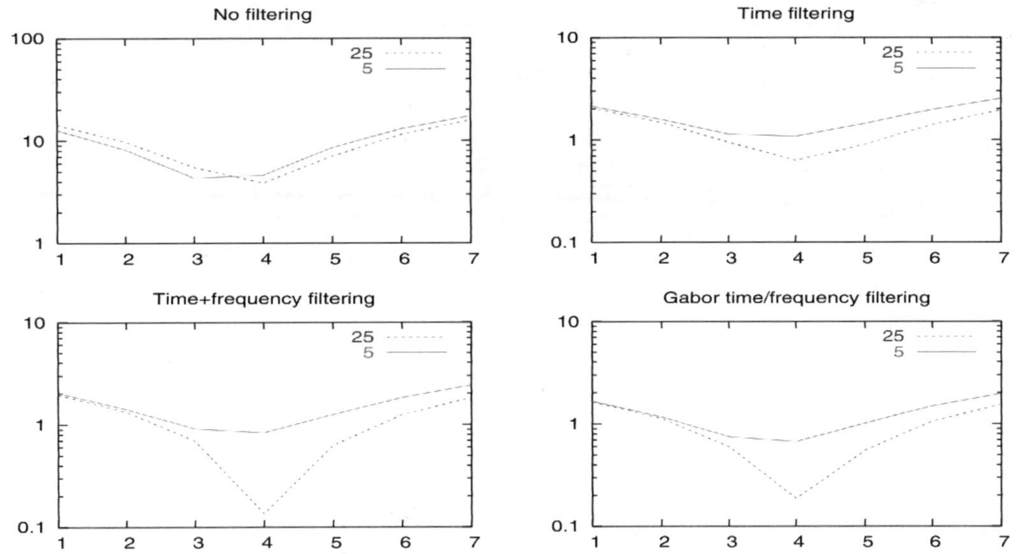

Figure 15. . Each plot shows one of the four norms around the best ($\Delta t = 25$) and second-best ($\Delta t = 5$) local minimum. The x axis ranges from 1 to 7 and the y axis ranges from 1 to 100 (upperleft) and 0.1 to 10 (other three plots) in semi-logarithmic plots.

5. Conclusion

The first subgroup on the Danfoss Flow Division problem dealt with obtaining a general model based on the physics of piezoceramic structures and matching layers. The model allows any time-dependent input to be analyzed using Fourier-transform techniques. A MatLab model is constructed for this purpose and some examples are considered. The second subgroup analyzed the influence of coding the input signal so as to obtain an unambiguous and accurate detection of arrival times based on traditional auto- and cross-correlation techniques. The main conclusion of this subgroup is that in the case with a set of narrowband transducers it is almost impossible to get improvements with a coded input signal since the response is almost exclusively at the centre frequency. Best compromise seems to be a dual frequency input 1 MHz +/- 5% with pair of transducers with centre frequencies at 1 MHz +/- 5% as well. Alternatively a swept frequency signal from 950 kHz to 1050 kHz may be used with the same set of transducers. The third subgroup has demonstrated on a real-world example that the new suggested methods work. However, since the examples given to us are few and the heuristics involved are many, we cannot say if these ideas can be used for production. Much further testing is needed: To determine the best choice of parameters (windows sizes and shapes, heuristics to find the beginning of the signal etc.), and also to determine which of the methods to use. The three subgroups have developed computer codes in MatLab used in the calculation of results presented (computer codes have been transferred to Danfoss A/S).

6. Acknowledgements

This study was partially supported by Danfoss A/S, Nordborg, Denmark.

7. References

[1] W. P. Mason, 'Electromechanical transducers and wave filters,' p. 205, D. Van Nostrand Company, Inc., Second Edition, 1948.

[2] M. Redwood, 'Transient performance of a piezoelectric transducer,' Journal of the Acoustical of America, Vol.33, p. 527–536, 1961.

[3] R. Krimholtz, D.A. Leedom, and G.L. Matthaei, 'New equivalent circuits for elementary piezoelectric transducers,' Electron. Lett. Vol.6, p. 398–399, 1970.

[4] E. K. Sittig, 'Transmission parameters of thickness-driven piezoelectric transducers arranged in multilayer configuration,' IEEE Trans. on Sonics and Ultrasonics, Vol. SU-14, p. 167-174, 1967.

[5] C. S. Desilets, J.D. Fraser, and G.S. Kino, 'The design of efficient broad-band piezoelectric transducers,' IEEE Trans. on Sonics and Ultrasonics, SU-25, p. 115-125, 1978.

[6] L. W. Briggs, 'The DFT, An Owner's Manual for the Discrete Fourier Transform,' SIAM, 1995.

[7] O. Christensen, 'An Introduction to Frames and Riesz Bases,' Birkhuser, 2003.

[8] K. Gröchenig, 'Foundations of Time-Frequency Analysis,' Birkhäuser, 2001.

[9] T. Strohmer, 'Numerical Algorithms for Discrete Gabor Expansions,' in *Gabor Analysis and Algorithms*, p. 267 - 294, Editors Feichtinger/Strohmer, Birkhaüser, 1998.

Institute of Physics Publishing
doi:10.1088/1742-6596/52/1/007
Journal of Physics: Conference Series **52** (2006) 73–82
Mathematics for Industry in Denmark

Exhaustive global grid search in computing receiver position from modular satellite range measurements

Niilo Sirola
Mathematics, Tampere University of Technology, PO Box 553, FI-33101 Tampere, Finland
E-mail: `niilo.sirola@tut.fi`

Abstract. A global grid search algorithm with an application in weak signal satellite positioning is implemented and tested numerically. The algorithm consists of local Gauss-Newton search and a global starting point chooser, and it can be generalized as a global optimization method for functions with an attraction basin of computable minimum radius. The method is shown to find the global minimum in a bounded region in predictable time.

1. Introduction

The Global Positioning System (GPS) is a satellite positioning system that was built for military purposes, but can be used for free by anyone with a suitable receiver [1]. Currently, the number of civil and commercial GPS users is growing faster than ever.

The GPS position is computed as a solution to a nonlinear optimization problem. The basic GPS algorithms have been used for decades and found to perform well. However, with GPS receivers becoming more common, there is a growing demand for precise positioning in environments very different to those for which the GPS system was originally designed. While the satellites and the signals they are broadcasting have remained essentially unchanged for the last three decades, the continuing development of receivers and navigation algorithms has enabled significant improvements in accuracy and availability.

Due to the low power of the satellite signals, the GPS receiver needs a direct line of sight with several satellites, which is easy in the air, open seas, and deserts but often impossible indoors, near tall buildings or under trees. The receiver has to decode several layers of data from the signal, which is not always possible or practical when the signal is attenuated and noisy [2].

In this paper, we study a case where only a part of the necessary information for GPS positioning is available, and the problem becomes a mixed integer nonlinear optimization problem. We eliminate the integer variables and devise a global optimization algorithm to solve the problem.

2. Background

Each GPS satellite transmits a signal that provides a means to determine the position of the satellite and the distance between the satellite and the receiver. The position of the receiver in n dimensions can be computed from n distance measurements.

We model the GPS positioning problem as the equation system

$$\rho_i = \|\mathbf{s}_i - \mathbf{r}\|_2 + \beta, \quad i = 1 \ldots n \tag{1}$$

where $\rho_i \in \mathbb{R}$ is called the pseudo-range. Pseudo-range is equal to the Euclidean distance between the ith satellite $\mathbf{s}_i \in \mathbb{R}^3$ and the receiver $\mathbf{r} \in \mathbb{R}^3$ plus an additive bias β that is common to all the pseudo-ranges. The bias results from the measurement process and is treated as the fourth unknown of the positioning problem. This model has been simplified by ignoring various errors and corresponding corrections that must be considered in actual GPS computations. These include the signal travelling in the atmosphere, Earth rotation, instability of satellite and receiver clocks, special relativity, and satellite orbit prediction errors, and their corrections. For a more comprehensive treatment on the GPS, see [1] for example.

Let the pseudoranges be measured such that $\hat{\rho}_i = \rho_i + \varepsilon_i$, where ε_i are independent measurement errors with zero mean. Given four pseudo-range measurements and the corresponding satellite positions, we can solve the system (1) and obtain estimates for the receiver position and bias. When $n > 4$, the system does not necessarily have an exact solution and a least-squares estimator is needed. Several closed-form [3, 4] as well as iterative [5, 6] estimators have been published for this problem.

The existence and uniqueness of the solution have been studied through geometric interpretation, and it has been found that in case $n = 4$, there may be two distinct solutions or no solutions at all [7]. For overdetermined cases, it is noted that a solution always exists in least squares sense and is unique "in all but degenerate cases", but the cost function is often bimodal, the other minimum fortunately always lying somewhere in outer space.

There are no convergence proofs published for the iterative methods, at least known to the author. The typically used Gauss-Newton iteration [8] seems to converge to the correct solution from practically any starting point on Earth, and it is widely agreed among GPS practitioners that the lack of convergence proofs is no problem.

3. Problem formulation

In this paper, we consider the case where the pseudo-ranges are measured modulo a constant Λ (≈ 300 km), and the satellite positions are known as functions of time but the exact time is not known. This corresponds to the situation where the satellite signals are very weak or are observed only for a short period [9].

We may write the pseudo-ranges in terms of whole and fractional multiples of Λ, such that $\rho_i = \Lambda N_i + \phi_i$ where N_i is a positive integer and $-\frac{1}{2}\Lambda \leq \phi_i < \frac{1}{2}\Lambda$. Now the system (1) becomes

$$N_i\Lambda + \phi_i = \|\mathbf{s}_i(t) - \mathbf{r}\|_2 + \beta, \quad i = 1 \ldots n \tag{2}$$

where $\mathbf{s}_i(t) \in \mathbb{R}^3$ is the position of ith satellite as the function of time. The functions $\mathbf{s}_i$ are assumed known and the code phases ϕ_i can be measured. The magnitude of the measurement errors is negligible when compared to the other quantities involved, and we develop the theory assuming errorless measurements.

The system (2) has unknowns t, $\mathbf{r}$, β and additionally the integers N_i, which we eliminate from the system rather than solving them explicitly. Define the lambda-fraction operator as

$$\mathrm{frac}_\Lambda(x) = x - \Lambda \cdot \mathrm{round}(\frac{x}{\Lambda}) \tag{3}$$

where round is the usual rounding to the nearest integer with half-integers rounded up.

Now denote $\mathbf{x} = [t \;\; \mathbf{r}^T \;\; \beta]^T \in \mathbb{R}^5$ and assume $n \geq 5$. Let $\mathbf{x}_\mathrm{t}$ be the true solution.

The system

$$\mathbf{q}(\mathbf{x}) = \begin{bmatrix} q_1(\mathbf{x}) \\ \vdots \\ q_n(\mathbf{x}) \end{bmatrix} = \begin{bmatrix} \mathrm{frac}_\Lambda(\|\mathbf{s}_1(t) - \mathbf{r}\|_2 + \beta - \phi_1) \\ \vdots \\ \mathrm{frac}_\Lambda(\|\mathbf{s}_n(t) - \mathbf{r}\|_2 + \beta - \phi_n) \end{bmatrix} = \mathbf{0}. \tag{4}$$

is equivalent to the system (2), and the minimizer of $\min \frac{1}{2}\|\mathbf{q}(\mathbf{x})\|_2^2$ is the best estimate for $\mathbf{x}_t$.

A global optimization method is called hybrid if it has a global stage to find a coarse approximation to the global minimum and a local search that is guaranteed to converge quickly when started sufficiently close to the optimum [10].

If the local search has an attraction basin with known minimum radius, the global search then only has to find at least one feasible point that is inside the attraction basin and then start the local search to yield the global minimum. This idea is the basis for the algorithm we propose to solve the positioning problem.

4. Local search

We seek the approximate solution for $\mathbf{q}(\mathbf{x}) = \mathbf{0}$ with the Gauss-Newton iteration [8, Section 8.5]

$$\mathbf{x}_{k+1} = \mathbf{x}_k - \left[J(\mathbf{x}_k)^T J(\mathbf{x}_k)\right]^{-1} J(\mathbf{x}_k)^T \mathbf{q}(\mathbf{x}_k) \tag{5}$$

where the Jacobian of $\mathbf{q}(\mathbf{x})$ is

$$J(\mathbf{x}) = \begin{bmatrix} \mathbf{u}_1^T \mathbf{s}_1'(t) & -\mathbf{u}_1^T & 1 \\ & \vdots & \\ \mathbf{u}_n^T \mathbf{s}_n'(t) & -\mathbf{u}_n^T & 1 \end{bmatrix}, \quad \mathbf{u}_i = \frac{\mathbf{s}_i(t) - \mathbf{r}}{\|\mathbf{s}_i(t) - \mathbf{r}\|_2}. \tag{6}$$

The iteration is terminated when $k > k_{\max}$ or $\|\mathbf{x}_{k+1} - \mathbf{x}_k\|_2 < \epsilon$, where $k_{\max}$ and ϵ are some predetermined values.

This iteration converges given a starting point $\mathbf{x}_0$ sufficiently close to the minimum when $\mathbf{q}''(\mathbf{x})$ is sufficiently small [11, Chapter 6]. The solution minimizes $\|\mathbf{q}(\mathbf{x})\|_2$. The rate of convergence, when convergence occurs, is no worse than linear.

Definition 1 *Let $\mathbf{x}^*$ be a solution to $\mathbf{f}(\mathbf{x}) = \mathbf{0}$. Let $B(\mathbf{x}^*, r) = \{\mathbf{x} \mid \|\mathbf{x} - \mathbf{x}^*\| < r\}$. If the Gauss-Newton iteration converges to $\mathbf{x}^*$ from any starting point inside an open ball $B(\mathbf{x}^*, r)$, then this ball is an **attraction basin** of $\mathbf{x}^*$.*

For the unknown vector $\mathbf{x} = [t \;\; \mathbf{r}^T \;\; \beta]^T \in \mathbb{R}^5$ in the system (4), we find it advantageous to use the following norm instead of the Euclidean one.

Definition 2 *Define **κ-norm** as*

$$\|\mathbf{x}\|_\kappa = \kappa\,|t| + \|\mathbf{r}\|_2 + |\beta| \tag{7}$$

where κ is a positive constant dependent on the relative geometry between the user and the satellite system:

$$\kappa = \max_{i,t,\mathbf{r}} \left| \mathbf{s}_i'(t)^T \frac{\mathbf{s}_i(t) - \mathbf{r}}{\|\mathbf{s}_i(t) - \mathbf{r}\|_2} \right| \tag{8}$$

Note that in the general case $\kappa = \max \|\mathbf{s}'(t)\|_2$. When the receiver is terrestrial, however, the velocities of the satellites are almost perpendicular to the line-of-sight vector $\mathbf{s}_i(t) - \mathbf{r}$ and κ is thus much smaller than the maximum velocity of the satellite. For example, for the GPS system and user restricted on the surface of the Earth, $\kappa \approx 830$ m/s.

We denote the open ball B_κ with

$$B_\kappa(\mathbf{p}, r) = \{\mathbf{x} \mid \|\mathbf{x} - \mathbf{p}\|_\kappa < r\}. \tag{9}$$

Next, we establish some properties of the system (4).

Lemma 1 *If* $\mathbf{x} \in B_\kappa(\mathbf{x}_t, \frac{\Lambda}{2})$, *then*

$$|q_i(\mathbf{x})| \le \|\mathbf{x} - \mathbf{x}_t\|_\kappa \tag{10}$$

Proof. Denote $\mathbf{x} = \mathbf{x}_t + \Delta\mathbf{x}$. Then consider the Taylor expansion

$$q_i(\mathbf{x}_t + \Delta\mathbf{x}) = \underbrace{q_i(\mathbf{x}_t)}_{=0} + q_i'(\mathbf{x}_t + \lambda\Delta\mathbf{x})\Delta\mathbf{x} \tag{11}$$

for some $\lambda \in (0,1)$. From

$$q_i'(\mathbf{x}) = \left[\frac{\mathbf{s}_i(t) - \mathbf{r}}{\|\mathbf{s}_i(t) - \mathbf{r}\|_2}^T \mathbf{s}_i'(t) \quad - \frac{\mathbf{s}_i(t) - \mathbf{r}}{\|\mathbf{s}_i(t) - \mathbf{r}\|_2}^T \quad 1 \right] \tag{12}$$

we notice that $q_i'(\mathbf{x}_t + \lambda\Delta\mathbf{x})$ is also of the form $[\mathbf{y}^T\mathbf{s}_i'(\tau)) \quad -\mathbf{y}^T \quad 1]$ where $\mathbf{y}$ is a unit vector. Thus

$$\begin{aligned} |q_i(\mathbf{x})| = |q_i(\mathbf{x}_t + \Delta\mathbf{x})| &= |q_i'(\mathbf{x} + \lambda\Delta\mathbf{x})\Delta\mathbf{x}| \\ &\le \left|\mathbf{y}^T\mathbf{s}_i'(\tau)\Delta t\right| + \left|\mathbf{y}^T\Delta\mathbf{r}\right| + |\Delta\beta| \\ &\le \kappa\,|\Delta t| + \|\Delta\mathbf{r}\|_2 + |\Delta\beta| = \|\mathbf{x} - \mathbf{x}_t\|_\kappa \end{aligned} \tag{13}$$

because, by definition (8), $\kappa \ge \left|\mathbf{y}^T\mathbf{s}_i'(\tau)\right|$. ■

Lemma 2 *The least-squares fit function* $f(\mathbf{x}) = \frac{1}{2}\|\mathbf{q}(\mathbf{x})\|_2^2$ *is smooth and has a unique minimum in* $\Omega = B_\kappa(\mathbf{x}_t, \frac{\Lambda}{2})$ *if* $\mathbf{q}'(\mathbf{x})$ *has full rank and* $\|(\mathbf{q}'(\mathbf{x})^T\mathbf{q}'(\mathbf{x}))^{-1}\sum_{i=1}^n q_i(\mathbf{x})q_i''(\mathbf{x})\| < 1$ *for all* $\mathbf{x} \in \Omega$.

Proof. If f is convex and has a critical point $\mathbf{x}^*$ in convex region Ω, then $\mathbf{x}^*$ is a global minimum of f. If f is strictly convex, the minimum is unique. [11, Theorem 9.4.1]

(i) The function f is smooth in Ω if $\mathbf{q}(\mathbf{x})$ is. The Euclidean norm as well as the satellite orbit functions are smooth, and lambda-fraction is smooth everywhere except when $|q_i(\mathbf{x})| = \Lambda/2$. But, when $\mathbf{x} \in \Omega = B_\kappa(\mathbf{x}_t, \frac{\Lambda}{2})$, then $|q_i(\mathbf{x})| < \Lambda/2$ by Lemma 1, and thus $\mathbf{q}$ is smooth in Ω.

(ii) Function f is strictly convex in Ω if f'' exists and is positive definite in Ω [8, 3.4.6].

$$f''(\mathbf{x}) = \mathbf{q}'(\mathbf{x})^T\mathbf{q}'(\mathbf{x}) + \sum_{i=1}^n q_i(\mathbf{x})q_i''(\mathbf{x}) \tag{14}$$

where the second term is very small. The first term is positive definite when $\mathbf{q}'(\mathbf{x})$ has full rank. Then $f''(\mathbf{x})$ is non-singular because $\|(\mathbf{q}'(\mathbf{x})^T\mathbf{q}'(\mathbf{x}))^{-1}\sum_{i=1}^n q_i(\mathbf{x})q_i''(\mathbf{x})\| < 1$ [12, Theorem 2.3.4]. Moreover, since the second term is symmetric and not large enough to reduce any of the eigenvalues of $f''(\mathbf{x})$ to zero, it cannot make any of them negative either, thus $f''(\mathbf{x})$ is positive definite.

(iii) The function f has a critical point $\mathbf{x}_t \in \Omega$ because $f'(\mathbf{x}_t) = \mathbf{q}'(\mathbf{x}_t)^T\mathbf{q}(\mathbf{x}_t) = \mathbf{q}'(\mathbf{x}_t)^T\mathbf{0} = \mathbf{0}$.

(iv) The region Ω is a unit ball and thus convex. ■

Although convexity or unique minimum do not guarantee the convergence of the Gauss-Newton iteration (5), numerical testing corroborates the conjecture that Ω is an attraction basin for $\mathbf{x}_t$. This is because the q_i are locally almost linear. The local search thus virtually guarantees good solution given a starting point $\mathbf{x}_0 \in \Omega$. Although there are several convergence theorems for Newton-like methods [13], none of them was found to guarantee convergence in whole Ω in this case, so we will have to settle on numerical evidence until further study.

5. Global search

We will now have to assume that the true solution $\mathbf{x}_t$ is the global minimum of f over the whole feasible region A. This is not true in the general case. Since each $q_i(\mathbf{x})$ has infinite number of roots, it is easy to construct satellite geometries where $f(\mathbf{x}) = 0$ for several $\mathbf{x}$. This happens also in the real world, but the probability is small at least when $n \geq 6$.

The role of the global stage of the algorithm is to find the feasible region $A \subset \mathbb{R}^5$, generate starting points in A, run the local search for each starting point, and determine which of the resulting local minima is the solution.

The global method could be a stochastic method such as Monte-Carlo or simulated annealing. In general case, these do not guarantee global convergence [10]. Instead, we develop a deterministic global stage feasible point finder that is guaranteed to find at least one point in an attraction basin of given size. First, define a covering.

Definition 3 *A point set P* ***covers*** *a region A with* ***covering radius*** *r if for any $\mathbf{x} \in A$ there is $\mathbf{p} \in P$ such that $\mathbf{x} \in B(\mathbf{p}, r)$.*

The set $\{B(\mathbf{p}, r) \mid \mathbf{p} \in P\}$ is called the ***covering*** *of A. It is called the* ***optimal covering*** *if there is no covering with smaller cardinality.*

It is relatively easy to find a covering, but the optimal coverings are known only for a few simple cases [14], and there are no general algorithms for finding the sharp covering radius of a given point set. Fortunately, we can quite easily generate a point set with covering radius at most a given r, a trivial example being a rectangular lattice with spacing $2r/\sqrt{p}$, where p is the number of dimensions.

Combining the concepts of attraction basin and covering radius gives the following convergence theorem for global optimization.

Theorem 1 *Let $f : A \subset \mathbb{R}^n \to \mathbb{R}$ have a global minimum $\mathbf{x}^*$. Let the open ball $B(\mathbf{x}^*, d)$ be an attraction basin of $\mathbf{x}^*$. Let P be a point set that covers A with covering radius smaller than d.*

Then the Gauss-Newton method converges to $\mathbf{x}^*$ from at least one point in P.

Proof. Because P has covering radius less than d in A, there exists by Definition 3 some $\mathbf{p} \in P$ such that $\|\mathbf{p} - \mathbf{x}^*\| < d$. Then $\mathbf{p} \in B(\mathbf{x}^*, d)$, which is an attraction basin for $\mathbf{x}^*$, and the Gauss-Newton iteration converges by Definition 1. ■

Note that if A is bounded, then P has a finite number of points, and the global minimum is found by starting the local search from each point and choosing the smallest of the resulting minima. This requires a total of at most $k_{\max} \cdot |P|$ Gauss-Newton iterations (5) and the basic algorithm is as follows.

Algorithm 0:

generate P that covers the feasible region A with radius smaller than d
for each $\mathbf{x} \in P$
  run the local search with starting point $\mathbf{x}$
  if the search converged, add the obtained minimum $\mathbf{x}^*$ to solution list X
end for
the global optimum is the local minimum with the smallest cost function value $f(\mathbf{x}^*)$

Returning to our particular problem, the feasible region A has several natural constraints. The system time t is restricted within the time of validity of the satellite orbit functions, which is about four hours. The receiver position $\mathbf{r}$ lies on or near the Earth surface, between some predetermined minimum and maximum altitudes. This forms a spherical shell. Finally, the bias can be restricted between $-\Lambda/2$ and $\Lambda/2$. Thus, the feasible region A is bounded and Algorithm 0 can be used.

We arrange the search grid as a Cartesian product of the time grid T, position grid R and bias grid B. The most natural time and bias grids are evenly spaced with ΔT and ΔB as the differences between elements. We construct the position search grid on the surface of the Earth analogous to a rectangular grid with spacing ΔR.

Taking into account that receiver position may be up to, say, 8 kilometers above the Earth, the point set $P = T \times R \times B$ then covers the feasible region A with covering radius at most $d = \frac{1}{2}\|[\Delta T\ \Delta R\ \Delta R\ 8\text{km}\ \Delta B]\|_\kappa$. This is not an optimal covering but fast to generate. The grid spacing parameters ΔT, ΔR, and ΔB can be freely chosen as long as $d < \Lambda/2$.

When using, for example, $\Delta T = 120$ s, $\Delta R = 100$ km, and $\Delta B = 60$ km, we get about 2.5 million points in P. Set $k_{\max} = 8$. Since one iteration of (2) takes about one millisecond (on a 800MHz Pentium III), the whole search would last less than six hours.

The computation can be made faster by restricting the feasible region and speeding the local search. In this paper we concentrate on restricting the feasible region.

Lemma 3 *If a terrestrial receiver sees satellites* $\mathbf{s}_1 \ldots \mathbf{s}_n$ *simultaneously, then the receiver position* $\mathbf{r}$ *and time* t *must satisfy the visibility conditions*

$$\mathbf{s}_i(t)^T\mathbf{r} \geq r_{\text{Earth}}^2, \quad i = 1 \ldots n \tag{15}$$

Proof. For a satellite $\mathbf{s}_i$ to lie over the horizon when viewed from point $\mathbf{r}$ on Earth, the line-of-sight vector $(\mathbf{s}_i - \mathbf{r})$ must lie on or above the local horizontal plane. The horizontal plane has normal vector parallel to $\mathbf{r}$, so the line-of-sight vector must satisfy $0 \leq \mathbf{r}^T(\mathbf{s}_i - \mathbf{r}) = \mathbf{r}^T\mathbf{s}_i - \|\mathbf{r}\|_2^2$. ∎

The visibility condition typically excludes over 90 percent of the search space. The more satellites, the smaller is the visible area on average.

Rather than checking each of the 2.5 million points against the visibility condition, it is faster to compute the visibility region explicitly at each t and cover only the visible area of Earth with R.

Furthermore, if we construct the search grid with covering radius d considerably smaller than $\Lambda/2$, we may reject the starting points with too large residuals because, by Lemma 1, $|q_i(\mathbf{x})| \leq \|\mathbf{x} - \mathbf{x}_t\|_\kappa < d$. These ideas give the following improved algorithm.

Algorithm 1:
choose a radius d and generate a search grid $T \times R \times B$ with covering radius at most d
for each $t \in T$
  if the visible region $V(t)$ is empty, proceed to next t
  generate R_t that covers $V(t)$
  for each $\mathbf{r} \in R_t$ and $b \in B$
    if for any i, $|q_i(\mathbf{x})| \geq d$, proceed to next point
    run the local search with starting point $\mathbf{x}$
    if the search did not converge in $k_{\max}$ iterations, proceed to next point
    if the altitude of the solution $\mathbf{x}^*$ is out of bounds, proceed to next point
    add $\mathbf{x}^*$ to solution list X
  end for
end for
if X is nonempty, choose the solution with smallest cost function value $f(\mathbf{x})$

How should we choose ΔT, ΔR, and ΔB for maximum efficiency? Large values minimize the number of points to search grid, but on the other hand the local search has to be started from all points. Choosing very small values produces very dense search grid, but most of the points can be rejected based on large residuals without having to start the local search. The optimal choice lies somewhere between the extremities and depends on the relative efficiently of the different parts of the implementation. One possibility is to use heuristic optimization with the computation time of a fixed set of positioning cases as the cost function to find the optimal parameters.

6. Numerical testing

Because of the lack of a theoretical proof of $\Omega = B_\kappa(\mathbf{x}_t, \frac{\Lambda}{2})$ being an attraction basin of the true solution, it is important to verify this assumption at least numerically. The model and algorithms were implemented in Matlab. Using actual GPS orbit data for satellite tracks, a million sets of pseudo-range measurements to 5–11 satellites were generated for randomly chosen true solutions. For each true solution, a random starting point was generated such that about half of the starting points lay inside the assumed basin of attraction Ω and about half outside, and the position estimate computed with the iteration (5).

The results, as listed in Table 1, show that the search converged from all starting points in Ω. Convergence occurred also from some points outside Ω, even as far as 2Λ away from the minimum. Inside Ω it took a maximum of five iterations for the search to converge and one iteration took less than one millisecond of computation time.

Table 1. Convergence tests from inside and outside attraction basin Ω

	$\mathbf{x}_0 \in \Omega$	$\mathbf{x}_0 \notin \Omega$
converged to $\mathbf{x}_t$	543686	155918
converged to other minimum	0	277814
diverged	0	22582

A similar simulation was also run with Gaussian noise added to the measurements. The noise had one sigma of 15 meters, which is a typical value for a GPS receiver in poor signal conditions. Although these errors are small when compared with the other quantities in the equations, they can cause very large errors in the position solution when the satellite geometry is near degenerate. Therefore an additional test for the condition of the approximate Hessian $J^T J$ (see Eq. 5) was added and iterations finishing with an ill-conditioned Hessian were considered divergent. As seen in Table 2, bad geometries happen also inside the basin of attraction, but still no false solutions are found.

Table 2. Convergence tests with noisy measurements

	$\mathbf{x}_0 \in \Omega$	$\mathbf{x}_0 \notin \Omega$
converged to $\mathbf{x}_t$	519050	142469
converged to other minimum	0	276288
diverged	24300	37893

The global algorithm could not be tested as extensively because of much longer computation times. Setting $\Delta T = 60$ s, $\Delta R = 80$ km, and $\Delta B = 33$ km reduced the execution time to about half when compared to the initial choices, but not much more tweaking was done on the Matlab code itself. However, the computation time could be drastically reduced by implementing the inner loops with a compiled language instead of Matlab. Preliminary results of this approach have been presented in a conference paper [15].

When there is no initial information about time t other than the four-hour period, a single position fix takes an average of 10 minutes to compute with Algorithm 1. The couple of hundred position fixes computed indicate that the correct solutions are found when $n \geq 7$, and most of the time when $n = 6$.

When an approximate time t_0 is known within one minute of the true time, the search is very fast because the solver can reduce the time search grid T to just one point. The simulation consisted of 10000 position fixes from random coordinates, of which about 93% produced the correct position estimate. As expected, all of the failed searches occurred when using six satellites. In average, one search took only a few seconds, the worst-case computation time being 31 seconds for six satellites and 4.8 seconds for eleven.

7. A numerical example

For a practical example with real-world data, we took a set of actual GPS measurements and chopped them to modulo Λ to get the measurement

$$\phi = [-99575 \quad 124457 \quad 3316 \quad 106688 \quad -60690 \quad -129977 \quad -17471 \quad 143623]^T \text{ m}.$$

These measurements were made on June 18th 2003 at about 12:12:46 UTC (GPS week 199, GPS time 303166 s) to GPS satellites 1, 25, 13, 4, 11, 30, 7, and 20, in that order.

In addition to the modular range measurements, the input information consists of the orbit functions for the eight satellites. The orbit functions actually used in GPS positioning have sixteen parameters, but for illustrative purposes we use just a four-parameter truncation: radius of the orbit r_i, inclination of orbit I_i, longtitude of the ascending node Ω_i, and argument of perigree ω_i. The position of ith satellite at time t can then be computed from

$$\mathbf{s}_i(t) = r_i \begin{bmatrix} \cos\phi_i \cos\Omega_i - \sin\phi_i \cos I_i \sin\Omega_i \\ \cos\phi_i \sin\Omega_i + \sin\phi(t) \cos I_i \cos\Omega_i \\ \sin\phi_i \sin I_i \end{bmatrix}, \quad \phi_i = \sqrt{\frac{\mu}{r_i^3}}t + \omega_i \tag{16}$$

where $\mu = 3.986005 \cdot 10^{14} \, \frac{\text{m}^3}{\text{s}^2}$ is the Earth's universal gravitational constant. The values of the satellite orbit parameters for the eight satellites are given in Table 3.

Table 3. Example satellite orbit parameters

i	r_i / m	I_i / rad	Ω_i / rad	ω_i / rad
1	26669227	0.97473256	2.75165527	0.52532410
2	26840334	0.94188151	-2.58537518	0.65523150
3	26506596	0.97894959	2.73303135	0.43003718
4	26567035	0.96493395	0.65864692	1.45288770
5	26604895	0.91225596	0.54902289	3.05310217
6	26385686	0.94217798	-1.48942712	1.98092506
7	26614305	0.93954499	-0.45657542	2.52472498
8	26499187	0.96450580	1.67606780	2.02825701

The time of validity for the satellite orbit functions is $t \in [296010, 310410]$ s.

We proceed according to Algorithm 1 by first choosing the search grid $\Delta T = 60$ s, $\Delta R = 80$ km, and $\Delta B = 33$ km.

The four-hour time search space is divided to 240 time steps, of which 145 have a non-empty visibility region, containing a total of 2326980 points. The condition $|q_i(\mathbf{x})| < d$ holds for 77421 points, and the local search is started from each one. 68049 of the searches converge to some local minimum within eight iterations, and after pruning out the solutions at too high or too low altitude, 2739 solution candidates remain and are added into the solution list.

The solution list entry with the smallest smallest objective function is chosen as the solution, and is in this case

$$t \approx 303165.9 \text{ s}, \quad \mathbf{r} \approx [2795073 \quad 1236237 \quad 5579648] \text{ m}, \quad \beta \approx -59556 \text{ m}$$

with the objective function value $f(\mathbf{x}) \approx 682.2$.

There are 22 other solution candidates within 10^{-4} seconds and 10 cm of this one, with objective function values less than 683. This is because the dense search grid contains multiple points within the attraction basin of the true solution. The next smallest objective function value, however, is over $2 \cdot 10^8$ and the corresponding position thousands of kilometers away.

In this experiment, the true position was known to be around [2795076 1236239 5579613], and the resulting positioning error 35 meters is well within the expected GPS accuracy. The search took little less than 14 minutes.

8. Conclusions

A solution method was developed for a special case of GPS positioning problem with minimal initial information required. The proposed method is applicable also to other satellite positioning systems, such as the European Galileo, and even to terrestrial positioning systems where distances are measured by repeating codes and positioning stations are relatively far away from the receiver.

Most of the previous GPS algorithms assume that the exact time can be deciphered from the signal and satellite positions can thus be accurately determined. The concept of time recovery, where the time has to be solved along with receiver position and bias was first suggested in [9] but with the requirement of an initial point sufficiently close to the solution. The modular GPS range problem has also been addressed by several authors, but the solutions assume that an initial position estimate within 150 km ($\approx \Lambda/2$) from the true position is known [6]. The present work gives bounds for a feasible initial estimate also in terms of time and bias, and provides an algorithm for positioning that requires the time to be known only within four hours.

Acknowledgments

This work was funded by Nokia Mobile Phones. The author would like to acknowledge Jari Syrjärinne at Nokia Mobile Phones for the original problem formulation and sharing his insight to the problem, and professor Robert Piché at Tampere University of Technology for his advice.

References

[1] Kaplan E D (ed.) 1996 *Understanding GPS: principles and applications* (Norwood: Artech House)

[2] Syrjärinne J 2001 *Studies of modern techniques for personal positioning* Dissertation, Tampere University of Technology

[3] Bancroft S 1986 An algebraic solution of the GPS equations *IEEE T. Aero. Elec. Sys.* **21** 56–9

[4] Leva J L 1996 An alternative closed-form solution to the GPS pseudo-range equations *IEEE T. Aero. Elec. Sys.* **32** 1430–9

[5] ICD-GPS-2000 2000 *NAVSTAR GPS space segment/navigation user interfaces, interface control document (ICD), Rev. C (Public Release Version)*, (Fountain Valley: ARINC Research Corporation)

[6] Dailey D J and Bell B M 1996 A method for GPS positioning, *IEEE T. Aero. Elec. Sys.* **32** 1148–54

[7] Abel J S and Chaffee J W 1991 Existence and uniqueness of GPS solutions *IEEE T. Aero. Elec. Sys.* **27** 952–6

[8] Ortega J M and Rheinboldt W C 1970 *Iterative solution of nonlinear equations in several variables* (Orlando: Academic Press)

[9] Syrjärinne J 2000 Time recovery through fusion of inaccurate network timing assistance with GPS measurements *Proc. 3rd International Conference on Information Fusion*, volume II, WeD5-3 – WeD5-10

[10] Xu P 2003 A hybrid global optimization method: The multi-dimensional case *J. Comput. Appl. Math.* **155** 423–46

[11] Fletcher R 1987 *Practical methods of optimization* (Chichester: John Wiley & Sons)

[12] Golub G H and Van Loan C F 1989 *Matrix Computations* 2nd edition (The Johns Hopkins University Press)

[13] Galántai A 2000 The theory of Newton's method *J. Comput. Appl. Math.* **124** 25–44

[14] Conway J H and Sloane N J A 1993 *Sphere packings, lattices and groups* 2nd ed (New York: Springer-Verlag)

[15] Sirola N and Syrjärinne J 2002 GPS position can be computed without the navigation data *Proc. ION GPS 2002 conference*, Portland, Oregon

Institute of Physics Publishing | Journal of Physics: Conference Series **52** (2006) 83–96
doi:10.1088/1742-6596/52/1/008 | Mathematics for Industry in Denmark

Mathematical analysis of the dynamic flow characteristic in a damping nozzle for a pressure transmitter

P. D. Howell
Mathematical Institute, 24–29 St Giles', Oxford OX1 3LB, UK
E-mail: howell@maths.ox.ac.uk

A. A. Lacey
School of Mathematical & Computer Sciences, Heriot–Watt University, Edinburgh EH14 4AS, UK
E-mail: a.a.lacey@hw.ac.uk

J. Vogler
Danfoss A/S, DK–6430 Nordborg, Denmark
E-mail: vogler@danfoss.com

A. Vonsild
Danfoss A/S, DK–6430 Nordborg, Denmark
E-mail: alv@danfoss.com

Abstract. We analyse the flow of a liquid along a pipe and through a small nozzle connecting the pipe to a pressure sensor. The aims are to predict the accuracy with which the pressure in the sensor reflects that in the pipe, and to determine whether or not cavitation occurs in the reservoir.

Keywords: Compressible pipe flow, cavitation.
MSC: 76N15

1. Introduction

Danfoss manufacture sensors that dynamically measure the pressure in liquid flowing along a pipe. The pressure in the pipe is transmitted via a small tube to a reservoir of liquid, whose pressure is measured using a piezo-electric membrane. In pipe flows where the liquid is stopped suddenly, it is quite possible for the pressure to become low enough for the liquid to cavitate. However, it is believed that the membrane may be damaged if cavitation bubbles ever form in the reservoir. Danfoss have discovered that this can be prevented by inserting a so-called "pulse snubber", which restricts the flow to a small hole between the pipe and the reservoir. It is not fully understood, however, how this protects the reservoir against cavitation, or how it affects the accuracy with which the pressure in the reservoir reflects that in the pipe. Our aim in this report is to construct mathematical models that address these questions.

We begin in section 2 by presenting a model for the flow in the pipe, incorporating liquid and vapour phases with a simple cavitation law governing the transition between them. In section 3, the model is used to predict the pressure transient caused when a flowing column of liquid is impulsively brought to rest. The resulting flow through the pulse snubber is then analysed in section 4. This is used in section 5 to predict the pressure in the reservoir corresponding to a given pipe pressure. Unfortunately, the model is unable to explain how the pulse snubber prevents cavitation in the reservoir. We hypothesise that this is a consequence of our over-simplified cavitation law, and propose two possible improvements in section 6. Finally, we draw our conclusions in section 7.

2. Model for flow in the pipe

Our aim in this report is to understand how pressure pulses in a pipe are transmitted to an attached pressure sensor. Before attempting to do so, we first consider the question of how the pressure pulses arise in the first place. This provides the input for our models of the pressure sensor and the pulse snubber developed below. Although we will only consider pipes with uniform cross section, we develop a model in which the pipe radius R and cross-sectional area $A = \pi R^2$ may both be functions of distance x along the pipe. This will prove useful later, since the same governing equations may then be applied to the flow through the pulse snubber.

Since we wish to consider situations in which the liquid may cavitate, we suppose in general that the pipe contains a mixture of gas and liquid. Conservation of mass implies that the cross-sectionally averaged density ρ and velocity u satisfy

$$\frac{\partial}{\partial t}(A\rho) + \frac{\partial}{\partial x}(A\rho u) = 0. \tag{1}$$

Conservation of momentum for the mixture leads to the equation

$$\frac{\partial}{\partial t}(A\rho u) + \frac{\partial}{\partial x}\left(A\rho u^2\right) = -A\frac{\partial p}{\partial x} - D, \tag{2}$$

where p is the average pressure and D is the drag exerted on the fluid by the pipe wall. In (2), we have approximated the profile coefficient as unity,[1] which is reasonable for turbulent flow [1].

Our constitutive equation for the wall drag takes the general form

$$D = -C 2\pi R\rho u|u|, \tag{3}$$

where C is a drag coefficient. In general, C may depend both on the flow regime (e.g. turbulent or laminar, single-phase or multiphase, annular or dispersed) and on the local Reynolds number,

$$\mathrm{R} = \frac{\rho u R}{\mu}, \tag{4}$$

where μ is the viscosity of the mixture. For single-phase turbulent flow, though, C typically depends only weakly on R; Blasius [2], for example, gives the empirical relation

$$C \approx \frac{0.033}{\mathrm{R}^{1/4}}. \tag{5}$$

[1] that is, we have approximated the average of the velocity squared by the square of the average velocity

For simplicity, we treat C as a constant (of the order of 0.01), so (2) reduces to

$$\frac{\partial u}{\partial t} + u\frac{\partial u}{\partial x} = -\frac{1}{\rho}\frac{\partial p}{\partial x} - \frac{2C}{R}u|u|. \tag{6}$$

To close the model, we need an effective gas law, relating pressure and density, for the mixture. Let the vapour volume fraction be α, so that $\alpha = 0$ in regions containing only liquid, $\alpha > 0$ where cavitation has occurred, and $\alpha = 1$ where there is only vapour. The density of the mixture is thus related to the liquid and vapour densities (ρ_l and ρ_v respectively) by

$$\rho = (1-\alpha)\rho_l + \alpha\rho_v. \tag{7}$$

We assume that the vapour phase satisfies an ideal gas law, with gas constant Γ, while the liquid phase has constant compressibility $1/\beta$. The two densities are thus related to the pressure by

$$p = \rho_v\Gamma T = p_a + \beta\left(\frac{\rho_l}{\rho_0} - 1\right), \tag{8}$$

where ρ_0 is the liquid density at atmospheric pressure p_a and T is the absolute temperature, assumed constant.

Equations (7) and (8) may be simplified by making the following observations.

(i) The vapour density ρ_v is much less than the liquid density ρ_l.

(ii) The incompressibility β is typically very large, so the liquid phase is effectively incompressible, with constant density $\rho_l \approx \rho_0$.

Thus the mixture density is well approximated by

$$\rho \approx \rho_0(1-\alpha), \tag{9}$$

and we need not solve for ρ_v.

The simplest cavitation model is that cavitation occurs whenever the pressure falls to the vapour pressure p_v, so that p is bounded below by p_v. This approach is common in the modelling of cavities [3] and may be expressed as the complementarity condition

$$p \geq p_v, \quad \alpha \geq 0, \quad \alpha(p - p_v) = 0. \tag{10}$$

Equations (1), (6), (9) and (10) form a closed system for ρ, u, α and p.

3. Impulsively stopped flow

Now we apply the model developed in section 2 to a typical experimental setup, in which liquid travelling along a uniform pipe (i.e. with constant A and R) is impulsively stopped by a valve. The situation is depicted schematically in figure 1. The liquid flows along the pipe at constant speed U such that, at $t = 0$, a column of length L has entered the pipe. The end is then closed instantaneously, but the liquid continues to flow, leaving behind a cavitated region. Our aim is to predict the pressure variations measured by a sensor placed a small distance downstream of the valve.

We denote the closed end by $x = 0$ and the boundary between the cavitated and non-cavitated regions by $x = X(t)$. The problems to be solved in the two regions are

(i) <u>$x < X(t)$, $\alpha > 0$, $p = p_v$</u>

$$-\frac{\partial \alpha}{\partial t} + \frac{\partial}{\partial x}\Big((1-\alpha)u\Big) = 0, \tag{11}$$

$$\frac{\partial u}{\partial t} + u\frac{\partial u}{\partial x} = -\frac{2C}{R}u|u|, \tag{12}$$

$$u = 0 \quad \text{at} \quad x = 0; \tag{13}$$

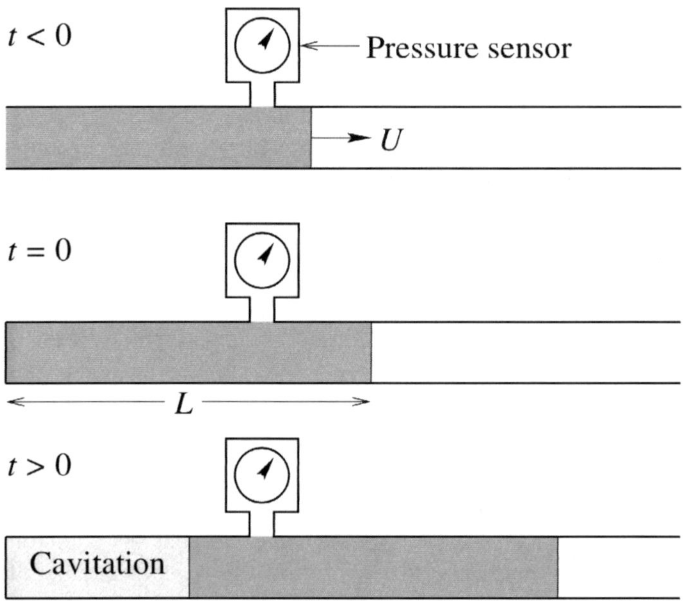

Figure 1. Schematic of impulsively stopped pipe flow.

(ii) $x > X(t)$, $\alpha = 0$, $p > p_v$

$$\frac{\partial u}{\partial x} = 0, \tag{14}$$

$$\frac{\partial u}{\partial t} + u\frac{\partial u}{\partial x} = -\frac{1}{\rho_0}\frac{\partial p}{\partial x} - \frac{2C}{R}u|u|, \tag{15}$$

$$u = U \quad \text{at} \quad t = 0. \tag{16}$$

In $x < X(t)$, we integrate the hyperbolic equation (12) to obtain

$$u = \frac{R}{2Ct}\left(1 - \mathrm{e}^{-2Cx/R}\right). \tag{17}$$

Then, (11) implies that

$$\frac{\mathrm{d}}{\mathrm{d}t}\left(t(\alpha - 1)\mathrm{e}^{-2Cx/R}\right) = 0 \tag{18}$$

along the characteristics $\mathrm{d}x/\mathrm{d}t = u$. The term in brackets is clearly zero when $t = 0$, and it follows that α must be equal to 1 for all positive t. The fluid in $x < X(t)$ is therefore entirely cavitated, with no liquid phase present at all. On the other hand, the vapour fraction α is zero in $x > X(t)$, so there must be a shock at $x = X(t)$. Conservation of mass and momentum across the shock leads to two Rankine–Hugoniot conditions, namely

$$\frac{\mathrm{d}X}{\mathrm{d}t} = \frac{\left[u(1-\alpha)\right]_-^+}{\left[1-\alpha\right]_-^+} = \frac{\left[\rho_0(1-\alpha)u^2 + p\right]_-^+}{\left[\rho_0(1-\alpha)u\right]_-^+}. \tag{19}$$

Now, in $x > X(t)$, (14) implies that u is a function only of t and, from (19) we deduce that

$$u(t) = \frac{\mathrm{d}X}{\mathrm{d}t}. \tag{20}$$

Then (15) gives the pressure gradient as

$$-\frac{1}{\rho_0}\frac{\partial p}{\partial x} = \frac{\mathrm{d}^2 X}{\mathrm{d}t^2} + \frac{2C}{R}\frac{\mathrm{d}X}{\mathrm{d}t}\left|\frac{\mathrm{d}X}{\mathrm{d}t}\right|. \tag{21}$$

The second Rankine-Hugoniot condition (19) implies that p is continuous across $x = X(t)$, while p is equal to atmospheric pressure p_a at the free end $x = X + L$:

$$p = p_v \text{ at } x = X(t), \quad p = p_a \text{ at } x = X(t) + L. \tag{22}$$

We assume here that the length L of the liquid column remains constant, which is consistent with our neglecting the mass of the cavitated gas. It follows that the pressure in the liquid is given by

$$p = p_v + \frac{p_a - p_v}{L}(x - X), \tag{23}$$

and (21) gives us a differential equation for $X(t)$:

$$\frac{\mathrm{d}^2 X}{\mathrm{d}t^2} + \frac{2C}{R}\frac{\mathrm{d}X}{\mathrm{d}t}\left|\frac{\mathrm{d}X}{\mathrm{d}t}\right| = -\frac{p_a - p_v}{\rho_0 L}. \tag{24}$$

The solution of (24) with initial conditions

$$X = 0, \quad \frac{\mathrm{d}X}{\mathrm{d}t} = U \quad \text{at } t = 0, \tag{25}$$

is

$$X = \frac{R}{2C}\log\left\{\cos(\lambda t) + \frac{\rho_0 U L \lambda}{p_a - p_v}\sin(\lambda t)\right\}, \quad \text{where } \lambda = \sqrt{\frac{2C(p_a - p_v)}{\rho_0 R L}}. \tag{26}$$

This solution is valid until $\dot{X}$ reaches zero, which happens when $t = t^*$, $X = X^*$, where

$$t^* = \frac{1}{\lambda}\tan^{-1}\left(\frac{\rho_0 U L \lambda}{p_a - p_v}\right), \quad X^* = \frac{R}{4C}\log\left(1 + \frac{2\rho_0 C L U^2}{R(p_a - p_v)}\right). \tag{27}$$

For $t > t^*$, $\dot{X}$ is negative, and the solution of (24) with $X = X^*$, $\dot{X} = 0$ at $t = t^*$ is

$$X = X^* - \frac{R}{2C}\log\left[\cosh\Big(\lambda(t - t^*)\Big)\right] = \frac{R}{4C}\log\left\{\frac{1 + \dfrac{2\rho_0 C L U^2}{R(p_a - p_v)}}{\cosh^2\Big(\lambda(t - t^*)\Big)}\right\}. \tag{28}$$

Hence X reaches zero, so the liquid plug hits the valve again, when

$$\cosh^2\Big(\lambda(t - t^*)\Big) = 1 + \frac{2\rho_0 C L U^2}{R(p_a - p_v)}. \tag{29}$$

When this happens, the liquid is moving at speed $u = -U^*$, where

$$U^* = \frac{U}{\sqrt{1 + \dfrac{2\rho_0 C L U^2}{R(p_a - p_v)}}}. \tag{30}$$

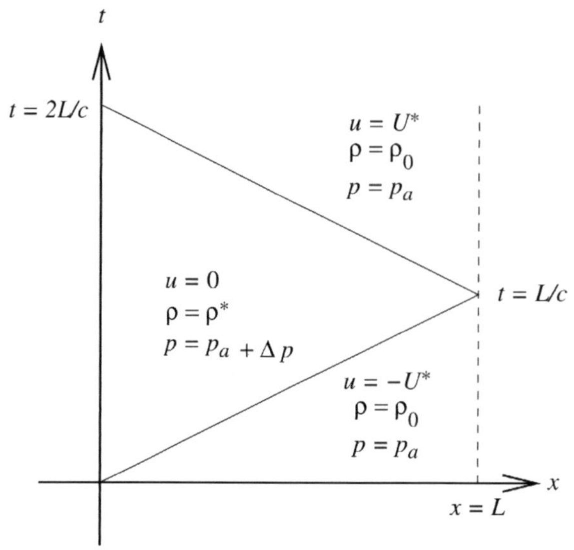

Figure 2. Sketch of the (x, t)-plane for the water rebound problem.

Note that the liquid hits the valve at a speed less than U because of the drag exerted during its journey up the pipe and back.

Now the liquid must bounce off the valve and set off up the pipe again. In the absence of any information about the elastic response of the valve, we make the simplest possible assumption that it is perfectly rigid. Then, to understand how the liquid manages to reverse direction, we have to include compressibility of the liquid, although this is assumed negligible elsewhere in our analysis. We can, however, still use the fact that the incompressibility β is large compared to any typical pressure in the problem, so the equations for conservation of mass and momentum may be linearised to

$$\frac{\partial \rho}{\partial t} + \rho_0 \frac{\partial u}{\partial x} = 0, \quad \rho_0 \frac{\partial u}{\partial t} = -\frac{\beta}{\rho_0}\frac{\partial \rho}{\partial x}. \tag{31}$$

It follows that u, ρ and p all satisfy the linear wave equation, with wave-speed c given by

$$c^2 = \frac{\beta}{\rho_0}. \tag{32}$$

Our linearisation of (31) is valid provided the flow is everywhere very subsonic, i.e. $|u| \ll c$.

The (x, t)-plane for the problem is sketched in figure 2. For simplicity we redefine the time origin such that, at $t = 0$, the water hits the valve at speed $u = -U^*$. Then its left-hand end $x = 0$ is reduced instantaneously to $u = 0$ and a pressure wave propagates into the water at speed c. Behind the wave, the water is stationary and the density is found by solving (31):

$$u = -U^* H(x - ct) \quad \Rightarrow \quad \rho = \rho_0 + \frac{\rho_0 U^*}{c} H(ct - x), \tag{33}$$

where H is the Heaviside function. Hence the density and pressure behind the wave are given by

$$\rho^* = \rho_0(1 + U^*/c), \quad p = p_a + \Delta p, \quad \Delta p = \frac{\beta U^*}{c}. \tag{34}$$

When the wave reaches the free boundary $x = L$, where p is fixed at atmospheric pressure p_a, it bounces back towards $x = 0$. Thus the pressure and density are returned to their atmospheric

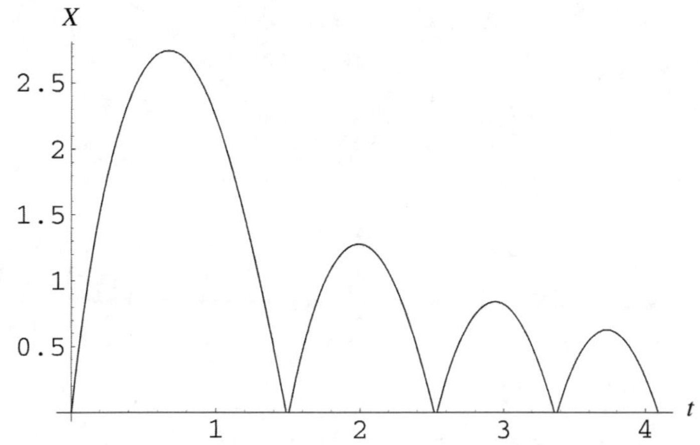

Figure 3. Free position X(m) versus time t(s) for a liquid plug bouncing up and down a pipe; parameter values given in equation (35).

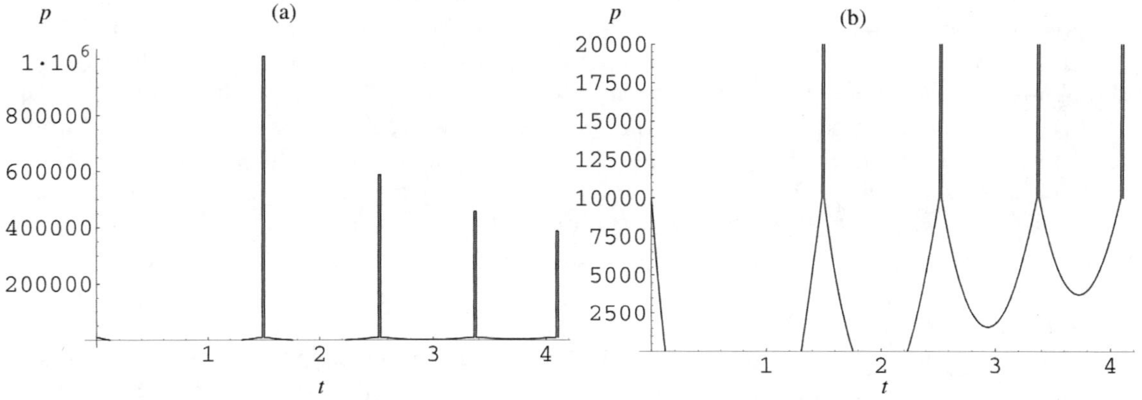

Figure 4. (a) Pressure p(Pa) at $x = 1\,\mathrm{m}$ versus time t(s). (b) Close-up of low-pressure behaviour. Parameter values given in equation (35).

values, while u ends up equal to $+U^*$. After a total time of $2L/c$, the whole liquid plug has reversed its direction of flow, and then sets off into $x > 0$. The whole process is then repeated, with the liquid now starting at speed U^* instead of U.

We illustrate the typical behaviour of $X(t)$ in figure 3, using the following estimated parameter values:

$$R = 1\,\mathrm{m}, \quad L = 10\,\mathrm{m}, \quad U = 10\,\mathrm{m\,s^{-1}},$$
$$\rho_0 = 10^3\,\mathrm{kg\,m^{-3}}, \quad p_a - p_v = 10^5\,\mathrm{Pa}, \quad \beta = 10^9\,\mathrm{Pa}, \quad C = 0.01. \tag{35}$$

The plug sets off from $X = 0$ at speed U and is decelerated both by the pressure difference $p_a - p_v$ applied across it and by the drag. It eventually turns around and returns to hit $x = 0$ at a speed slightly less than U. After a very short delay $2L/c = 0.02\,\mathrm{s}$, while the acoustic waves propagate through the liquid, it sets off again at the reduced speed and the whole process repeats as a series of gradually shortening excursions.

In figure 4 we show the corresponding pressure p measured at a sensor placed a distance $x = 1\,\mathrm{m}$ from the valve. Graph (b) shows how the pressure decreases as the liquid plug flows away from the valve, reaching the vapour pressure (effectively zero) when the cavitation bubble

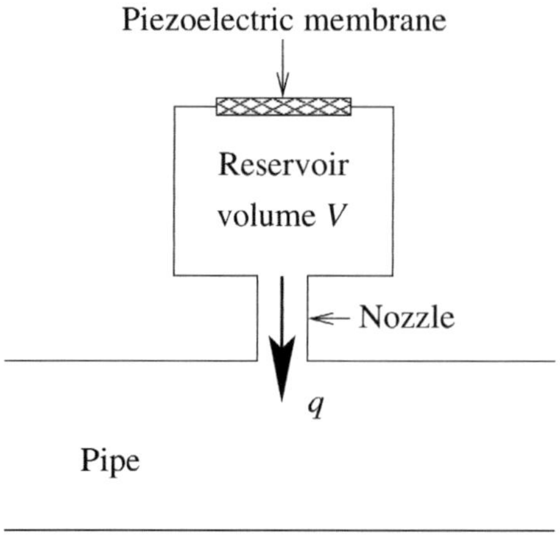

Figure 5. Schematic of a pressure sensor.

reaches the sensor. When the liquid returns and passes the sensor once more, the pressure starts to increase again. Then, when the liquid impacts with the valve, a very large pressure pulse Δp occurs, as given in (34). The pressure peak is too large to be shown in graph (b) so its full magnitude is indicated in graph (a). Then the liquid once again propagates away from the valve and the process repeats. Notice that, after the first two oscillations, the cavitation bubble never reaches the sensor so the pressure is uniformly positive. This behaviour is qualitatively similar to that reported by Danfoss, although the pressure peaks seem to be more smeared out in practice. This may be due to imperfect rigidity of the valve, attenuation of the signal before it reaches the sensor, or some other damping in the system.

4. Flow through the pulse snubber

In section 3, we found a typical pressure profile to be expected in a pipe following an impulsive shutdown of the flow. In practice, this would be measured by a sensor similar to that shown schematically in figure 5. A reservoir of volume V is connected to the pipe by a nozzle, and the pressure in the reservoir is measured by a piezoelectric membrane. The accuracy of the sensor relies on the pressure in the reservoir being close to that in the pipe. Furthermore, the piezoelectric membrane, although well able to withstand very high pressures, is thought to be susceptible to rupturing if the fluid in the reservoir cavitates. Our task now is, therefore, given a pressure profile in the pipe similar to that shown in figure 4, to predict the corresponding reservoir pressure.

The flow between the pipe and the reservoir is limited by the introduction of a "pulse snubber". This blocks the nozzle except for a very small hole through which fluid can propagate, whose typical profile is illustrated in figure 6. As a first approximation, we model the flow through this hole using the same equations as those developed for flow in the pipe in section 2. This enables us to estimate roughly how the flux through the hole depends on the material parameters and the geometry, although CFD calculations might be used in future to obtain more detailed and accurate results.

Let x and t denote distance along the hole and time respectively. As before, the gas fraction is denoted by α and the fluid velocity and pressure by u and p. Then, with the nozzle radius R a given function of x (similar to that illustrated in figure 6) and the density approximated by $\rho \approx \rho_0(1-\alpha)$, the equations governing conservation of mass and momentum are

$$-R^2\frac{\partial \alpha}{\partial t} + \frac{\partial}{\partial x}\Big((1-\alpha)R^2 u\Big) \quad = \quad 0, \qquad (36)$$

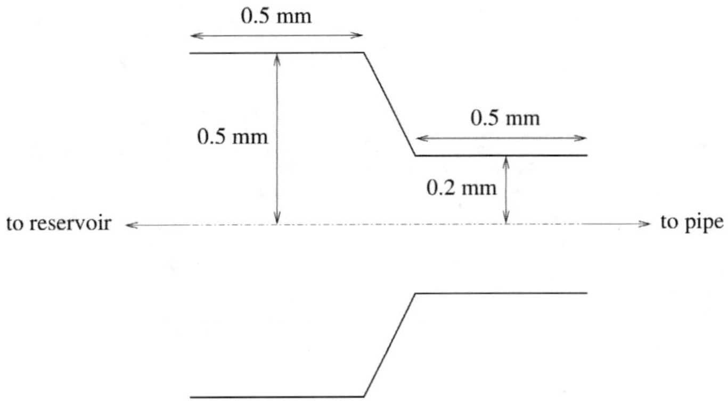

Figure 6. Schematic sketch of the cross-section through the hole in a pulse snubber.

$$\frac{\partial u}{\partial t} + u\frac{\partial u}{\partial x} = -\frac{1}{\rho_0(1-\alpha)}\frac{\partial p}{\partial x} - \frac{2C}{R}u|u|. \tag{37}$$

The problem is closed by our simple cavitation law

$$p \geq p_v, \quad \alpha \geq 0, \quad \alpha(p - p_v) = 0. \tag{38}$$

We assume that the flow through the hole is *quasi-steady*, so the time-derivatives in (36) and (37) may be neglected; this assumption will be justified later in section 5. Thus (36) implies that the mass flux q from the reservoir to the pipe is spatially uniform:

$$\pi\rho_0(1-\alpha)R^2u = q(t). \tag{39}$$

The momentum equation (37) takes two different forms depending on whether or not the liquid has cavitated:

$$\alpha = 0,\ p > p_v \quad \Rightarrow \quad \frac{\mathrm{d}p}{\mathrm{d}x} = \frac{2q}{\pi^2\rho_0 R^5}\left(q\frac{\mathrm{d}R}{\mathrm{d}x} - C|q|\right), \tag{40}$$

$$\alpha > 0,\ p = p_v \quad \Rightarrow \quad \frac{\mathrm{d}\alpha}{\mathrm{d}x} = \frac{2(1-\alpha)}{qR}\left(q\frac{\mathrm{d}R}{\mathrm{d}x} - C|q|\right). \tag{41}$$

These are to be solved subject to specified reservoir pressure $p_R(t)$ and pipe pressure $p_P(t)$. In matching these with the pressure in the hole, we also include the dynamic term in the Bernoulli constant:

$$p + \frac{1}{2}\rho_0 u^2 = p_R(t) \text{ at } x = 0, \quad p + \frac{1}{2}\rho_0 u^2 = p_P(t) \text{ at } x = L, \tag{42}$$

where L is the length of the hole. Our aim is to determine the flux q in terms of the pressure difference $\Delta p = p_R - p_P$.

We have to decide which of (40) and (41) to apply. The possibilities are: either the fluid cavitates nowhere in the pulse-snubber ($\alpha = 0$, $p > p_v$), or it cavitates everywhere ($\alpha > 0$, $p = p_v$), or cavitation occurs at some value of x, where $\alpha = 0$ and $p = p_v$.

Suppose first that q is positive, which will occur if $\Delta p > 0$. Then, since R is a decreasing function of x, (40) and (41) imply that p and α are both decreasing functions of x. It follows that cavitation cannot occur anywhere within the pulse snubber. If it did, then we would have $\alpha = 0$ and $p = p_v$ at (say) $x = x^*$ and thus either α or $p - p_v$ would have to be negative in $x > x^*$, neither of which is allowed. Moreover, if α were positive throughout, both p_R and p_P

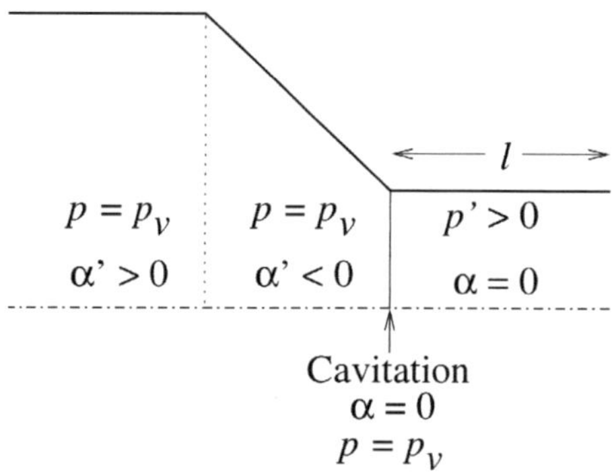

Figure 7. Schematic sketch showing the behaviour of the pressure p and vapour fraction α when the flux q is negative.

would have to be smaller than p_v. Hence (40) applies throughout the pulse snubber, and the pressure difference is thus given by

$$\Delta p = \frac{2Cq|q|}{\pi^2 \rho_0} \int_0^L \frac{\mathrm{d}x}{R^5}. \tag{43}$$

If there is no cavitation in the hole, then (43) still holds if q is negative. In this case, however, there is no guarantee that cavitation does not occur; the configuration shown schematically in figure 7 (where prime is used as shorthand for $\mathrm{d}/\mathrm{d}x$) is also possible. The pressure drop required to force fluid in through the narrow end reduces p to the vapour pressure at the point where the nozzle widens. The fluid is then cavitated throughout the wider region. If the length of the narrow section is denoted by l, then the pressure drop in this case is related to the (negative) flux q by

$$p_P - p_v = \frac{q^2}{2\pi^2 \rho_0 R(L)^5}\Big(4Cl + R(L)\Big). \tag{44}$$

This implies that equilibrium between p_R and p_P may be attained very rapidly when q is negative, since $p_P - p_v$ is usually very much larger than $p_P - p_R$.

In summary, we have obtained a law relating the flux q through the pulse snubber to the pressure drop applied across it. Depending on the geometry of the hole, we may have to apply the two different relations (43) for $q > 0$ and (44) for $q < 0$. Henceforth, we assume for simplicity that this is not the case so that (43) may be applied for all values of q.

Finally, we note the analogy between the analysis carried out here and the classical theory of compressible flow through a nozzle [5]. If our cavitation law (38) were replaced by a continuous functional relation between p and α, then (40) and (41) would become

$$\left(1 - \frac{c^2}{u^2}\right)\frac{\mathrm{d}\alpha}{\mathrm{d}x} = \frac{2(1-\alpha)}{R}\left(\frac{\mathrm{d}R}{\mathrm{d}x} - C\mathrm{sgn}(q)\right), \tag{45}$$

where c is the speed of sound:

$$c^2 = -\frac{1}{\rho_0}\frac{\mathrm{d}p}{\mathrm{d}\alpha}. \tag{46}$$

The switch between $p = p_v$ and $\alpha = 0$ is replaced in this regularised approach by a switch between supersonic ($u > c$) and subsonic ($u < c$) flow, which can only happen where $\mathrm{d}R/\mathrm{d}x - C\mathrm{sgn}(q)$ changes sign.

5. Model for the reservoir

In the previous section we derived a relation between the pressure difference $\Delta p = p_R - p_P$ and the flux q through the pulse snubber. The pipe pressure p_P is assumed to be a known function of time, but the reservoir pressure p_R must be found by analysing the behaviour of the fluid in the reservoir. We assume that the fluid in the reservoir is well-mixed, with density $\rho(t)$. If the reservoir volume is V, then simple mass balance gives

$$\frac{\mathrm{d}}{\mathrm{d}t}(\rho V) = -q, \tag{47}$$

where q is given by (43):

$$q = \pi\sqrt{\frac{\rho_0}{2CI}}\operatorname{sgn}(\Delta p)\sqrt{|\Delta p|}. \tag{48}$$

The problem is closed by applying the equation of state between p_R and ρ. If the fluid in the reservoir is uncavitated ($\alpha = 0$), then

$$p_R = p_a + \beta\left(\frac{\rho}{\rho_0} - 1\right). \tag{49}$$

Combining (47)–(49), we obtain a single differential equation for $p_R(t)$:

$$\frac{\mathrm{d}p_R}{\mathrm{d}t} = -\frac{\pi\beta}{V\sqrt{2CI\rho_0}}\operatorname{sgn}(p_R - p_P)\sqrt{|p_R - p_P|}. \tag{50}$$

Now we compare the left- and right-hand sides of (50) using the following estimates:

$$\begin{gathered} p_R \approx 10^6\,\mathrm{Pa}, \quad \mathrm{d}p_R/\mathrm{d}t \approx 10^8\,\mathrm{Pa\,s^{-1}}, \quad \beta \approx 10^9\,\mathrm{Pa}, \\ V \approx 1.5\times 10^{-6}\,\mathrm{m}^3, \quad C \approx 0.01, \quad I \approx 10^{15}\,\mathrm{m}^{-4}, \quad \rho_0 \approx 10^3\,\mathrm{kg\,m^{-3}}. \end{gathered} \tag{51}$$

These imply that the right-hand side of (50) is of order $10^9\,\mathrm{Pa\,s^{-1}}$ and therefore dominates the left-hand side. It follows that p_R and p_P are approximately equal, and the first correction may be obtained by substituting $p_R = p_P$ into the left-hand side of (50):

$$p_R \sim p_P - \frac{2CI\rho_0 V^2}{\pi^2\beta^2}\frac{\mathrm{d}p_P}{\mathrm{d}t}\left|\frac{\mathrm{d}p_P}{\mathrm{d}t}\right|. \tag{52}$$

The second term on the right-hand side of (52) indicates how the pressure in the reservoir lags behind that in the pipe. The fact that it is typically small compared with p_P shows that the measured pressure p_R is indeed close to the pipe pressure. However, the estimate (52) fails when the pipe pressure is constant, in particular when the fluid in the pipe has cavitated so that $p_P \equiv p_v$.

To examine the behaviour of p_R when this happens, suppose that the pipe pressure takes the form

$$p_P = \begin{cases} p_v - \phi t & t < 0 \\ p_v & t \geq 0 \end{cases} \tag{53}$$

where ϕ represents the rate at which p_P drops to p_v. For $t < 0$, the reservoir pressure is given approximately by

$$p_R \sim p_v - \phi t + \frac{2V^2 CI\rho_0\phi^2}{\pi^2\beta^2}. \tag{54}$$

For $t > 0$, we have to solve

$$\frac{\mathrm{d}p_R}{\mathrm{d}t} = -\frac{\pi\beta}{V\sqrt{2CI\rho_0}}\sqrt{p_R - p_v}, \tag{55}$$

with

$$p_R = p_v + \frac{2V^2CI\rho_0\phi^2}{\pi^2\beta^2} \quad \text{at } t = 0. \tag{56}$$

The solution is

$$p_R = p_V + \frac{\pi^2}{\beta^2}8V^2CI\rho_0\left(\frac{4V^2CI\rho_0\phi}{\pi^2\beta^2} - t\right)^2, \tag{57}$$

so the reservoir pressure reaches the cavitation pressure after a finite time

$$t_v = \frac{4V^2CI\rho_0\phi}{\pi^2\beta^2}. \tag{58}$$

We can therefore determine whether or not the reservoir pressure has time to reach the cavitation pressure by comparing t_v with the time for which p_P remains at p_v. Using the estimates given in (51), we find that

$$t_v \approx 10^{-6}\,\mathrm{s}, \tag{59}$$

so that there will certainly be time for the reservoir pressure to reach p_v. This seems to imply that the liquid in the reservoir is likely to cavitate. To escape this conclusion, a more complicated cavitation model is required, so that the pressure falling to p_v does not necessarily lead to cavitation. We briefly discuss some possible approaches in section 6.

Before doing so, we use the estimates obtained in this section to justify the quasi-steady approximation adopted in section 4. The flux implied by (48) and (52) is of order

$$q \sim \frac{\rho_0 V}{\beta}\frac{\mathrm{d}p_p}{\mathrm{d}t}. \tag{60}$$

The velocity is thus of order

$$u \sim \frac{q}{\pi R^2\rho_0} \sim \frac{V}{\pi R^2\beta}\frac{\mathrm{d}p_p}{\mathrm{d}t}. \tag{61}$$

The time derivatives in (36) and (37) may be neglected provided $L \ll ut$, where t is a typical timescale. Thus we obtain the following condition for the quasi-steady approximation to apply:

$$\frac{\pi R^2 L\beta}{V\Delta p_P} \ll 1, \tag{62}$$

where Δp_P is a typical variation in p_P. This has the interpretation that the volume of the nozzle must be much less than the volume of fluid displaced through it. The estimates given in (51) imply that the left-hand side of (62) is small but not tiny; maybe of order 0.2. Hence the quasi-steady approximation is acceptable as a first approximation but should be examined in future refinements of the model.

6. Alternative cavitation models

In this report we have restricted our attention to the simple cavitation model (10). This assumes that the fluid cavitates whenever the pressure drops to the vapour pressure p_v. Unfortunately, this assumption leads to the conclusion that the liquid in the reservoir is almost certain to cavitate under realistic pipe pressures, despite the presence of the pulse snubber, which is not observed in practice. Furthermore, it is well known that liquids can exist at pressures well below the vapour pressure; indeed negative pressures of minus several hundred atmospheres

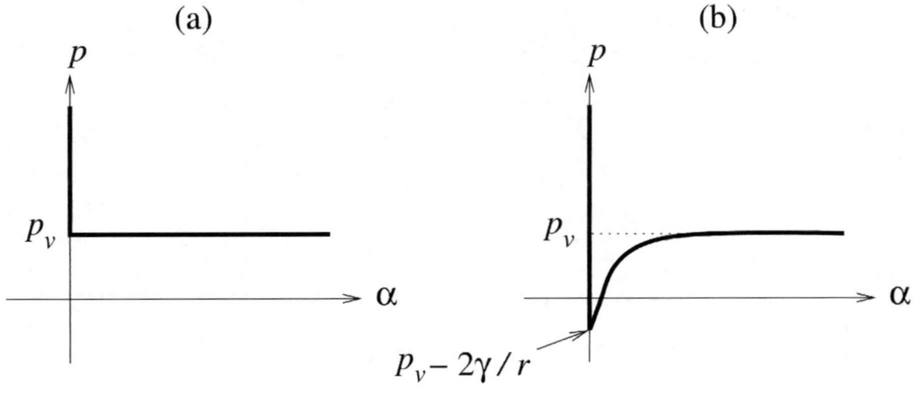

Figure 8. Possible behaviours of pressure versus vapour fraction.

have been measured under laboratory conditions [4]. Recent experiments carried out by Danfoss indicate that the fluid in the reservoir may remain in the liquid state although the pressure drops significantly below the vapour pressure. To explain this, a more complicated cavitation model is required, and we now describe briefly two possible approaches.

Cavitation is typically initiated via nucleation of tiny bubbles. A bubble of radius a containing vapour at pressure p_v will expand if the external pressure is less than $p_v - 2\gamma/a$, where γ is the surface tension of the vapour-liquid interface. It thus appears that an infinite negative pressure is required to make a bubble grow from an initially zero radius. The generally-accepted resolution of this paradox is that bubbles are formed at pre-existing nucleation sites with small but nonzero radius r. This implies that the liquid pressure must fall to $p_v - 2\gamma/r$ before cavitation occurs. To describe such behaviour, one might replace the degenerate p versus α behaviour depicted in figure 8(a) with something like that shown in figure 8(b). Here, the pressure must overcome the negative "barrier" $p_v - 2\gamma/r$ before α can increase from zero. Notice, though, that the problem is ill-posed in the region where $\mathrm{d}p/\mathrm{d}\alpha$ is positive, since the speed of sound is imaginary. Some sort of regularisation (such as surface tension) would be required to avoid a catastrophic instability.

An alternative approach is to consider the kinetics of the transition from the liquid to vapour state. A simple evolution equation for this transition is of the form

$$\tau \frac{\mathrm{d}\alpha}{\mathrm{d}t} = f(\alpha)\left(\frac{p}{p_v} - 1\right), \qquad (63)$$

where τ is the associated timescale and $f(\alpha)$ has zeros at the equilibria of the system, namely $\alpha = 0$ and $\alpha = 1$. If $f(\alpha)$ has the form sketched in figure 9, then the system equilibrates to $\alpha = 0$ (uncavitated) when $p > p_v$ or $\alpha = 1$ (cavitated) when $p < p_v$. Such a model therefore allows the possibility that the liquid in the reservoir may resist cavitation even though the reservoir pressure drops to p_v.

7. Conclusions

In this report we have constructed models for the flow of liquid along a pipe, through the pulse snubber connecting the pipe to a reservoir and in the reservoir itself. We allowed for cavitation by including a vapour volume fraction α that varies between 0 for pure liquid to 1 for pure vapour. Our basic model consists of two partial differential equations, representing conservation of mass and momentum, for α, the velocity u and the pressure p. To close the system, a cavitation law relating α to p is required, and we proposed the simple complementarity condition (10).

As a first application of our model, we considered the impulsively stopped flow of liquid along a pipe. The predicted damped oscillatory flow and corresponding pressure profile depicted

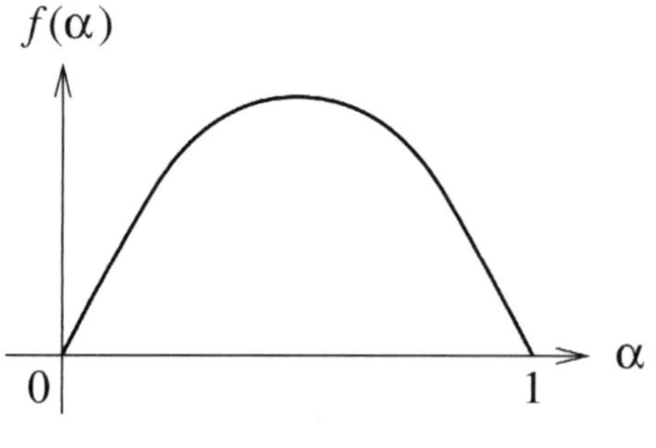

Figure 9. Sketch of the function $f(\alpha)$.

in figures 3 and 4 were found to show encouraging qualitative agreement with experimental measurements. We then used the same model to derive a relation between the pressure difference across the pulse snubber and the flux through it. We thus obtained the differential equation (50) for the reservoir pressure $p_R(t)$. This enabled us to predict the discrepancy between the reservoir and pipe pressures and to estimate whether or not the reservoir pressure is likely to fall to the vapour pressure.

There are many further avenues to be explored, including the following.

- The pressure peaks shown in figure 4 are significantly sharper than those reported in practice. Better predictions might be obtained by allowing for damped elastic response of the valve as the liquid plug hits it.
- We showed that it is impossible for the liquid to cavitate in the pulse snubber when the flow is from the reservoir to the pipe but, depending on the geometry, cavitation may occur when liquid flows from the pipe into the reservoir. This switch in behaviour warrants further study and may help to explain why the pulse snubber protects the reservoir from low pressures but not from high pressures.
- Our quasi-one-dimensional model for the flow through the pulse snubber cannot give more than order-of-magnitude estimates, since the geometry is far from slowly-varying. CFD calculations would enable more accurate predictions to be made, including time-dependent effects that we neglected.
- We proposed two possible improved cavitation laws in section 6, and further consideration of the underlying physics may yield more improvements. The effect of these on our predictions should be explored.
- In practice, the reservoir may be filled with air when it is first fitted. Our modelling approach should allow the time taken to expel all the air to be estimated.

Acknowledgements

We would like to thank the organisers and sponsors of the 47th European Study Group with Industry at the University of Southern Denmark. The input of manpower and technical information from Danfoss A/S is also gratefully acknowledged.

[1] A.C. Fowler, *Mathematical Models in the Applied Sciences*, Cambridge University Press, Cambridge, 1997.

[2] H. Blasius, Das Ähnlichkeitsgesetz bei Reibungsvorgängen in Flüssigkeiten, *Mitt. Forsch. Arb.* **131**, 1–39 (1913).

[3] G. Birkhoff and E.H. Zarantonello, *Jets, Wakes, and Cavities*, Academic Press Inc., New York, 1957.

[4] A.T.J. Hayward, Negative pressure in liquids: can it be harnessed to serve man? *American Scientist* **59**, 434–443 (1971).

[5] H. Ockendon and A.B. Tayler, *Inviscid fluid flows*, Springer-Verlag, New York, 1983.

AUTHOR INDEX